Valuing Ground Water

Economic Concepts and Approaches

Committee on Valuing Ground Water

Water Science and Technology Board

Commission on Geosciences, Environment, and Resources

National Research Council

NATIONAL ACADEMY PRESS
Washington, D.C. 1997

NATIONAL ACADEMY PRESS • 2101 Constitution Avenue, NW • Washington, DC 20418

NOTICE: The project that is the subject of this report was approved by the Governing Board of the National Research Council, whose members are drawn from the councils of the National Academy of Sciences, the National Academy of Engineering, and the Institute of Medicine. The members of the committee responsible for the report were chosen for their special competences and with regard for appropriate balance.

This report has been reviewed by a group other than the authors according to procedures approved by a Report Review Committee consisting of members of the National Academy of Sciences, the National Academy of Engineering, and the Institute of Medicine.

Support for this project was provided by the U.S. Department of Energy, U.S. Environmental Protection Agency Grant No. C-R-823279-01-3, U.S. Department of Defense/Defense Supply Service Grant No. DASW01-95-M-6159, and the National Water Research Institute.

Library of Congress Cataloging-in-Publication Data

Valuing ground water : economic concepts and approaches / Committee on Valuing Ground Water, Water Science and Technology Board, Commission on Geosciences, Environment, and Resources, National Research Council.

p. cm.

Includes bibliographical references and index.

ISBN 0-309-05640-3

1. Groundwater—Valuation. I. National Research Council (U.S.). Committee on Valuing Ground Water.

HD1691.V35 1997

333.91′04—dc21 97-4837

Valuing Ground Water: Economic Concepts and Approaches is available from the National Academy Press, 2101 Constitution Ave., NW, Box 285, Washington, DC 204185 (1-800-824-6242; http://www.nap.edu).

Cover art by Y. David Chung. Chung is a graduate of the Corcoran School of Art in Washington, D.C. He has exhibited his work throughout the country, including the Whitney Museum in New York, the Washington Project for the Arts in Washington, D.C., and the Williams College Museum of Art in Williamstown, Massachusetts.

Printed in the United States of America

COMMITTEE ON VALUING GROUND WATER

LARRY W. CANTER, *Chair*, University of Oklahoma, Norman
CHARLES W. ABDALLA, Pennsylvania State University, University Park
RICHARD M. ADAMS, Oregon State University, Corvallis
J. DAVID AIKEN, University of Nebraska, Lincoln
SANDRA O. ARCHIBALD, Hubert H. Humphrey Institute of Public Affairs, Minneapolis, Minnesota
SUSAN CAPALBO, Montana State University, Bozeman
PATRICK A. DOMENICO, Texas A&M University, College Station (*from September 1994 to November 1995*)
PETER G. HUBBELL, Water Resources Associates, Inc., Tampa, Florida
KATHARINE L. JACOBS, Arizona Department of Water Resources, Tucson
AARON MILLS, University of Virginia, Charlottesville
WILLIAM R. MILLS, JR., Orange County Water District, Fountain Valley, California
PAUL ROBERTS, Stanford University, California
THOMAS C. SCHELLING, University of Maryland, College Park
THEODORE TOMASI, University of Delaware, Newark

WSTB Liaison

HENRY J. VAUX, JR., University of California, Riverside

Staff

STEPHEN D. PARKER, Study Director *(September 1994 through January 1996)*
SHEILA D. DAVID, Study Director *(January 1996 through April 1997)*
ETAN GUMERMAN, Project Coordinator *(September 1994 through October 1996)*
MARY BETH MORRIS, Senior Project Assistant *(September 1994 through July 1996*)
ELLEN A. DE GUZMAN, Project Assistant *(July 1996 through April 1997)*

Consultant

JOEL DARMSTADTER, Resources for the Future *(September 1994 through June 1996*)

COMMISSION ON GEOSCIENCES, ENVIRONMENT, AND RESOURCES

The National Academy of Sciences is a private, nonprofit, self-perpetuating society of distinguished scholars engaged in scientific and engineering research, dedicated to the furtherance of science and technology and to their use for the general welfare. Upon the authority of the charter granted to it by the Congress in 1863, the Academy has a mandate that requires it to advise the federal government on scientific and technical matters. Dr. Bruce M. Alberts is president of the National Academy of Sciences.

The National Academy of Engineering was established in 1964, under the charter of the National Academy of Sciences, as a parallel organization of outstanding engineers. It is autonomous in its administration and in the selection of its members, sharing with the National Academy of Sciences the responsibility for advising the federal government. The National Academy of Engineering also sponsors engineering programs aimed at meeting national needs, encourages education and research, and recognizes the superior achievements of engineers. Dr. William A. Wulf is president of the National Academy of Engineering.

The Institute of Medicine was established in 1970 by the National Academy of Sciences to secure the services of eminent members of appropriate professions in the examination of policy matters pertaining to the health of the public. The Institute acts under the responsibility given to the National Academy of Sciences by its congressional charter to be an adviser to the federal government and, upon its own initiative, to identify issues of medical care, research, and education. Dr. Kenneth I. Shine is president of the Institute of Medicine.

The National Research Council was organized by the National Academy of Sciences in 1916 to associate the broad community of science and technology with the Academy's purposes of furthering knowledge and advising the federal government. Functioning in accordance with general policies determined by the Academy, the Council has become the principal operating agency of both the National Academy of Sciences and the National Academy of Engineering in providing services to the government, the public, and the scientific and engineering communities. The Council is administered jointly by both Academies and the Institute of Medicine. Dr. Bruce M. Alberts and Dr. William A. Wulf are chairman and vice chairman, respectively, of the National Research Council.

Preface

Ground water, while providing much of the nation's supplies of water for domestic, industrial, and agricultural purposes, is surprisingly underappreciated and usually undervalued. Water managers at various levels of government are faced with an array of decisions involving development, protection, and/or remediation of ground water resources. Examples of questions basic to such decisions at the local level include:

(1) Should ground water be used singly or in conjunction with surface water supplies to meet increasing water usage requirements?

(2) Should a comprehensive water conservation program be implemented in order to extend the availability of ground water and minimize or preclude ground water depletion?

Examples of questions basic to decisions at the state or federal level include:

(1) Are the benefits of ground water protection programs greater than their costs, and how should such wellhead protection efforts be funded?

(2) How should ground water remediation projects be prioritized given that the costs of remedial actions typically far exceed available funding? Should the value of ground water resources be considered in deciding if remediation efforts should be undertaken at a site?

Valuation of ground water resources is critical in determining an efficient outcome in each of these examples as well as many other ground water development, protection, and/or remediation projects, programs, or policy decisions. However, the ground water resource, a non-market good, is difficult to value;

and, as a result, economic valuation and future considerations have historically played almost no part in decision making.

The fundamental need to value natural resources was recognized in a 1990 report of the Science Advisory Board (SAB) of the U.S. Environmental Protection Agency (EPA, 1990). Based on the review of comparative risk assessments of environmental problems, a committee of 39 distinguished scientists, engineers, and other experts drawn from academia, state government, industry, and public interest groups developed ten recommendations; of relevance to this report is Recommendation 10—EPA should develop improved methods to value natural resources and to account for long-term environmental effects in its economic analyses (EPA, 1990).

In 1994 the U.S. Environmental Protection Agency requested that the National Research Council (NRC) appoint a committee to study approaches to assessing the future economic value of ground water, and the economic impact of the contamination or depletion of these resources. This committee was appointed in 1994 under the auspices of the NRC's Water Science and Technology Board. The committee was charged to:

(1) review and critique various approaches for estimating the future value of uncontaminated ground water in both practice and in theory;

(2) identify areas in which existing approaches require further development and promising new approaches which might be developed;

(3) delineate the circumstances under which various approaches would be preferred in practice for various applications of decision making regarding long-term resource use and management;

(4) outline legislative and policy considerations in connection with the use and implementation of recommended approaches, and related research needs; and

(5) illustrate, through real or hypothetical case examples, how recommended procedures would be applied in practice for representative applications.

Due to the relevance of the committee charge to other public interest groups and agencies, three other sponsors provided financial support for this NRC study in addition to EPA: the National Water Research Institute, the U.S. Department of Defense, and the U.S. Department of Energy.

The focus of the study on ground water valuation and the composition of the committee established the need for economists to work with ground water experts. Disciplines represented on the committee included agricultural economics, environmental engineering, hydrogeology, microbiology, public policy, resource economics, and water law. The members were drawn from academia, private consultants, and water management positions in local government.

While the assignment was challenging, the committee quickly agreed on three matters that provided its starting points. First, an interdisciplinary approach is necessary for ground water valuation studies. Second, when valuing ground water, the *in situ* and ecological services must be recognized along with the more obvious extractive services. Finally, it was recognized that common terminology was not available as a foundation for this study. Thus concepts and principles from environmental economics and ground water management had to be appropriately integrated to provide a basis for the work of the committee.

The committee has completed its task and, in so doing, has received considerable assistance from the NRC staff. Accordingly, on behalf of the committee, I wish to express our thanks to the following persons: Sheila David, Study Director; Etan Gumerman, Project Coordinator; Mary Beth Morris, Project Assistant; Ellen de Guzman, Project Assistant; Joel Darmstadter, Consultant; and Steve Parker, Director of the Water Science and Technology Board (WSTB). In addition, Henry Vaux, WSTB member and liaison to this committee, provided both helpful guidance and technical input.

Finally, I wish to express my appreciation to all committee members for their willingness to discuss new concepts from an interdisciplinary perspective, to prepare and revise materials for this report, and to strive for consensus-building on key issues. We have all learned from this process!

Larry W. Canter,
Chairman

Reference

U.S. EPA. 1990. Reducing Risk: Setting Priorities and Strategies for Environmental Protection, Science Advisory Board, Relative Risk Reduction Strategies Committee, U.S. EPA, Washington, D.C.

Contents

Valuing Ground Water

Executive Summary

Ground water in the United States is usually considered as either an invaluable good or as a "free" good. At one extreme, the Comprehensive Environmental Response, Compensation, and Liability Act (CERCLA or Superfund) implies a very high value for ground water by requiring restoration of contaminated water sources to drinking water quality. Billions of dollars have been spent to clean up contaminated ground water with little comparison of costs or technological difficulty to future benefits. At sites where cleanup is technically infeasible, the Superfund law essentially assigns an infinite value to the resource.

At the other extreme, historically, ground water has been priced well below its value and, as a consequence, misallocated. In many states and localities, no charge is imposed for water withdrawn, and the consumer, whether a public water supply entity, an individual, or a firm regards the cost as being confined to the energy used for pumping and the amortization of well construction and the costs of the treatment and distribution system. As a result, depletion and pollution continue largely because it is not recognized that ground water has a high or long-term value. Further, the Environmental Protection Agency's Science Advisory Board (SAB) report *Reducing Risk* (1990) has been perceived as not properly valuing ground water. The report neglects the uniqueness of the ground water resource and the often irreversible nature of ground water depletion and pollution, implying that declines in ground water quality and quantity need not be major concerns.

Such undervaluation of ground water fosters misallocation of resources in two ways: (1) the ground water resource is not efficiently allocated relative to alternative current and future uses; and (2) authorities responsible for resource management and protection devote inadequate attention and funding to maintaining ground water quality.

In 1994, recognizing the need for better methods and informed decision-making in this area, the U.S. Environmental Protection Agency, the National Water Research Institute, the U.S. Department of Energy, and the U.S. Department of Defense requested that the National Research Council undertake this study. This study examines approaches to assessing the future economic value of ground water as well as the economic impact of contaminating or depleting this resource. Key points addressed include the minimal historical attention given to ground water valuation in general, and specific methods that can be used to perform such valuation studies.

Until the last few decades, attention, even in natural resource and environmental economics, has been given primarily to the effects of exploiting natural resource assets such as extractive minerals, land and timber, ocean fisheries, and surface water resources. The economic value of unique natural and environmental resources, such as wetlands and other ecosystems, has more recently been considered. Most ground water studies to date have focused only on the valuation of limited production-related services provided by ground water, and not on a more comprehensive view of production and ecological services.

A fundamental step in valuing a ground water resource is recognizing and quantifying the resource's total economic value (TEV). Knowing the resource's TEV is crucial for determining the net benefits of policies and management actions. For purposes of this study, ground water services have been divided into two basic categories: extractive services and *in situ* services. Each of these has an economic value, and these values can be summed to yield TEV as follows:

TEV = extractive value + *in situ* value

The most familiar of these two components are the extractive values, which are derived from the municipal, industrial, commercial, and agricultural demands met by ground water. The *in situ* services (i.e., services or values that occur or exist as a consequence of water remaining in place within the aquifer) include, for example, the capacity of ground water to (1) buffer against periodic shortages in surface water supplies; (2) prevent or minimize subsidence of the land surface from ground water withdrawals; (3) protect against sea water intrusion; (4) protect water quality by maintaining the capacity to dilute and assimilate ground water contaminants; (5) facilitate habitat and ecological diversity; and (6) provide discharge to support recreational activities. The committee's calculation of TEV as the sum of extractive and *in situ* values can also be expressed by using concepts which often appear in the environmental economics literature. The relationship between those concepts and the ones in this report has been defined in Chapter 1. The committee developed the taxonomy in Chapter 1 so that its use will lead to greater potential for interdisciplinary work on the neglected service areas.

It is important to recognize the TEV of ground water even when one cannot develop specific quantitative separations of the various components. In fact, delineations of what can and cannot be quantified can be useful both to decision-makers for either development or remediation projects, and to researchers seeking to advance conceptual and methodological approaches. Descriptive information or surrogate quantitative measures that are not monetized may be the only information that can be assembled on some TEV components.

In many circumstances even a partial or inexact measurement of TEV can greatly aid decision-making by providing insight into how TEV changes with a policy or management decision. In some cases, the measurement of use values alone, or extractive services alone, can reveal substantial information on how the resource's TEV would be affected by a policy decision. In other circumstances, these limited measures may fail if they provide only a small portion of the components of TEV that would be altered.

GROUND WATER RESOURCES: HYDROLOGY, ECOLOGY, AND ECONOMICS

Valuation of the extractive and *in situ* services of ground water requires an understanding of the hydrology and ecology of the ground water source. Hydrologic information includes numerous factors such as rainfall, runoff, infiltration, and water balance data; depth to ground water; whether the water-bearing zone is confined or unconfined; ground water flow rates and direction; and type of vadose and water-bearing zone materials. The contribution of ground water to stream base flow and the relationships between ground water and wetland and lake ecosystems are also important.

Knowing natural recharge rates and spatial locations, along with ground water usage rates and trends, is also necessary in water balance calculations and the consideration of ground water depletion. Depending upon the location, relationships between sea water or saline water intrusion and ground water use may also need to be established. Land subsidence can occur in some areas if ground water use is excessive, causing major problems with infrastructure components such as building foundations, roads, sewers, and water and utility lines. The effect of subsidence on flooding (especially) in coastal areas may also be significant. All these should be considered in valuing a ground water resource.

Some ground water supplies can be viewed as nonrenewable because of the long time-frame required to replenish them. Depletion of ground water (including overdrafting and mining) in deep aquifers, for instance, is essentially irreversible. Therefore, because ground water is a unique and potentially exhaustible resource vital to future generations, the costs of valuation studies may be recovered by assisting in the protection of ground water. Without planning and protection of ground water, the resource may not be available to support future generations.

In other circumstances ground water overdraft can be economically efficient and socially beneficial in the short term. For all aquifers, a "steady state" should eventually be reached in which withdrawals are limited to recharge. The level at which this steady state is to be maintained is a matter of choice. During times of drought when surface supplies are scarce, temporary overdraft may be justified, with a subsequent reduction in use of the aquifer to let it recharge. The level would then fluctuate around some average steady state condition.

The tendency for ground water to be treated as an "open access" resource when it is exploited underscores the importance of well-defined, clearly enforceable rights to extract or obligations to protect ground water. In instances where these rights are not defined and enforceable, the availability of ground water is subject to the "law of capture," in which whoever gets to the water first gets first rights to it. If ground water is subject to the law of capture, then the benefits of protection, remediation, and enhancement investments will also be subject to the law of capture. This results in less than optimal investment in the preservation and enhancement of ground water quality, since those investing in such measures cannot reap all of the benefits. (Associated legal and institutional questions are discussed beginning on page 10.)

Treating environmental systems as economic assets that provide goods and services has become an established approach in environmental economics. Ground water systems create ecological services by providing discharge for the maintenance of stream flows and to wetlands and lakes. These discharges support general ecological functions that provide their own services of economic value. For example, discharge to aquatic ecosystems may aid preservation of threatened or endangered species and support downstream uses of water for drinking or irrigation. (Many flowing streams in the southwest U.S., for example, have gone dry after nearby aquifers were drawn on too heavily.) Ground water provides a "derived" value through its contributions to the larger environment.

While the valuation of a given ground water resource may be complex, several simple principles may be applied to almost any valuation problem:

- **Because ground water resources are finite, decision-makers should take a long-term view in all decisions regarding valuation and use of these resources, proceeding very cautiously with any actions that would lead to an irreversible situation regarding ground water use and management. Ground water depletion, for instance, is often irreversible. Some aquifers do not recharge quickly. Moreover, overdrafting can sometimes lead to a collapse of the geologic formation, permanently reducing the aquifer's storage capacity.**

- **Decision-makers should also be cautious regarding contamination of ground water. Restoration of contaminated aquifers, even when feasible, is resource-intensive and time-consuming. Restoration methods are uncertain and unlikely to improve significantly in the near future. As a result, it is**

almost always less expensive to prevent ground water contamination than to clean it up.

- **Ground water often makes significant contributions to valuable ecological services. For example, in the Southwest, many flowing streams have been eliminated by overpumping. Because the ground water processes that affect ecosystems and base stream flow are not well understood, combined hydrologic/ecologic research should be pursued to clarify these connections and better define the extent to which changes in ground water quality or quantity contribute to the change in ecologic values.**
- **Ground water management entities should consider appropriate policies such as pump taxes or quotas to ensure that the cost of using the water now rather than later is accurately accounted for by competing pumpers.**

VALUATION FRAMEWORK

One of the major challenges in valuing ground water is how to integrate the hydrologic and physical components of ground water resources into a valuation scheme. An appropriate conceptual basis for valuation identifies service flows as the central link between economic valuation and ground water quality and quantity.

Every generation should be concerned about the supply and quality of fresh water, and about who has access to it, at what cost. Defining the best long-term management of the resource requires balancing the needs of the present with those of the future. In theory, the balancing is done everyday by markets as reflected in the discount rate. However, many citizens, policy-makers, and scientists believe that the discount rate does not adequately consider the value of goods or services for future generations.

Discounting is a procedure that adjusts for future values of a particular good by accounting for time preferences. Higher discount rates, which give less weight to future net benefits, encourage present use and deter present investments. The market rate of interest will also influence individual and corporate decisions regarding resource extraction. Public entities can choose the discount rate they prefer, and much has been written about these choices. The discount rate a water utility employs when valuing ground water reflects perceptions of risk, returns, and possibly intergenerational equity. A high discount rate implicitly places a low value on the water's value to future generations. A low rate implies the opposite.

A valuation framework must take into account how time, institutions, water quality and quantity, hydrologic factors, and services interact to affect the resource's value. This necessity has several important implications:

- **As noted earlier, some knowledge of a resource's TEV is vital to the work of water managers, and in the development of policies dealing with**

allocation of ground water and surface water resources. For many purposes, the full TEV need not be measured, but in all cases where a substantial portion of the TEV will be altered by a decision or policy, that portion should be measured.

- **Policy-makers must recognize the impact that a utility's choice of a discount rate can have on ground water management decisions. Ideally, the discount rate should give adequate weight to long-term considerations.**
- **An interdisciplinary approach, such as the conceptual model presented in Chapter 3, is useful in conducting a ground water value assessment. The approach should incorporate knowledge from the economic, hydrologic, health, and other social, biological, and physical sciences. Assessments should be site specific and integrate information on water demands with information on recharge and other hydrologic concerns, and to the extent possible should reflect the uncertainties in both the economic estimates of the demand for ground water and in the hydrologic and biophysical relationships.**

VALUATION METHODS

Ground water services are difficult to value because much of the information needed for valuation is not readily available. Market trades can provide data useful in valuation, for instance, but most of the services provided by ground water are not traded on markets. However, techniques do exist for valuing nonmarket goods.

Economic value is not a fixed, inherent attribute of a good or service but rather depends on time, circumstances, and individual preferences. The economic value of a good or service can be inferred either from someone's willingness to pay (WTP) or willingness to accept compensation (WTA) for giving it up.

Several taxonomies have been developed to categorize the types of economic values associated with natural resources, such as a ground water system. One taxonomy distinguishes between use values, which are determined by the contribution of a resource to current or future production and consumption, and nonuse values, which typically refers to aesthetic or contemplative values arising from goods and services. The critical distinction for decision-making is between goods and services whose economic values are fully captured in market prices and those whose value is not thus captured.

Applicability of Valuation Methods

One prominent technique that attempts to measure total value, including use and nonuse values, is the contingent valuation method (CVM). CVM values are elicited directly from individuals (via interviews or questionnaires (see Appendix B)) in the form of statements of maximum WTP or minimum WTA compensation for hypothetical changes in environmental goods, such as ground water quantity or quality. The CVM can be applied to both ground water use and nonuse values. There are numerous methodological controversies associated with application of CVM, including how the hypothetical ground water change that people are being asked to value is to be specified, the elicitation format for asking valuation questions, the appropriate measure to be elicited (i.e., WTP or WTA), and various types of response biases.

The advantage of the contingent valuation method, however, is that it allows analysts to focus precisely on the total resource attribute (e.g., quantity or quality changes) to be valued. CVM provides reliable estimates of value when an individual has a close connection to the resource being valued. When there is a large nonuse component to the TEV being elicited, application of CVM is difficult, making it one of the most controversial areas in the valuation literature. CVM practitioners believe that it is the only method capable of capturing a substantial part of value when nonuse value is a large part of the TEV. However, the continuing controversy over both the theoretical validity and the practicality of CVM-based studies of nonuse values raises questions regarding its use in natural resource damage assessments and litigation situations. Table 1.6 in Chapter 1 and Table 4.5 in Chapter 4 compare the advantages and disadvantages of CVM along with other valuation methods.

In contrast to direct elicitation via CVM or some other stated preference technique, economists also have developed indirect methods (e.g., hedonic price models), which infer values from other behaviors associated with the good. A strength of indirect methods is that they rely on observed behaviors of producers and consumers. Examples of observed behaviors, such as how much water is applied in irrigation or as drinking water at a given cost, expenditures on water purification systems, or how much people will spend to travel to a recreational resource, help to establish a water resource's value. However, because indirect approaches generally measure only one component of the TEV (use value) and in some cases require large amounts of data, care must be taken when employing them.

In any case, for valid and reliable results to be obtained, the valuation method must be well-matched to the context and the ground water function/service of interest. (Chapter 4, Table 4.5 provides a summary of potential matches.) Methods for valuing the quality of drinking water include cost of illness, averting behavior, contingent valuation, and conjoint analysis (e.g., contingent ranking or behavior).

Uncertainty

The decision-maker attempting to value ground water faces significant uncertainties regarding hydrologic, institutional, economic, and human health aspects of ground water management. One source of uncertainty lies with the problem of predicting the consequences of environmental policies and actions. A related set of challenges stems from the difficulty of assessing ground water benefits in the future and the irreversible nature of some present ground water management decisions and impacts. Economic uncertainties regarding nonmarket goods and services are even more substantial because there is no accurate documentation of monetary values when markets are absent.

The notion of risk contrasts with uncertainty. Risk characterizes situations about which there are a known set of probabilities. By contrast, uncertainty characterizes situations in which the probabilities are incompletely known or unknown. Techniques of risk analysis can be customarily applied to characterize risky situations analytically. One method of accounting for risk involves addition of "risk premiums" to the discount rate. The size of the "risk premium" varies directly with the degree of risk. The concept of risk is extremely important in analyzing the potential costs associated with degraded water quality.

A careful consideration of these valuation factors leads to several conclusions:

- **For valid and reliable results to be obtained, the valuation method must be well-matched to the context and the ground water function or service of interest.**
- **It is hard to make generalizations about the validity and reliability of specific valuation approaches in the abstract. The validity of the approach depends on the valuation context and the type of ground water services that are of interest. Different approaches are needed to value different services; care must be taken not to double-count values associated with different services.**
- **Previous ground water valuation studies have focused primarily on a small part of the known ground water functions and services (identified in Chapter 3). Thus, the current empirical knowledge of the values of ground water is quite limited and concentrated in a few areas, such as extractive values related to drinking water use.**
- **The contingent valuation method (CVM), when used correctly, has the potential for producing reliable estimates of ground water use values in certain contexts. CVM has the advantage of allowing analysts to focus precisely on the total value of a resource attribute, compared to the results from other indirect approaches that generally fail to capture total economic value. However, few, if any, studies to date meet the stringent conditions, as established by a NOAA panel of Nobel-Laureate economists, that are required to produce defensible estimates of nonuse values. More research is needed to compare use values from CVM with those of other methods to determine whether CVM will consistently yield reliable estimates.**
- **Given the problems in using CVM to measure ground water values, EPA and other appropriate government agencies should encourage ways of enhancing the utility of CVM. For example, contingent ranking or behavior**

methods may be useful in improving the robustness of CVM estimates and may expand the potential for transferring existing CVM estimates to other empirical settings.

- **If data are available and critical assumptions are met, indirect valuation methods (e.g., travel cost method (TCM), hedonic price method (HPM), averting behavior) can produce reliable estimates of the use value of ground water.**
- **The EPA, and other federal agencies as appropriate, should develop and test other valuation methods for addressing the use and nonuse values of ground water, especially the ecological services provided by ground water.**
- **Technical, economic, and institutional uncertainties should be considered and their potential influence delineated in ground water valuation studies. Research is needed to articulate such uncertainties and their potential influence on valuation study results.**
- **Ground water values obtained from both indirect and direct methods are dependent on the specific ground water management context. Attempts to generalize about or transfer values from one context to another should be pursued with caution.**
- **If data are available and critical assumptions are accurate, traditional valuation methods such as cost of illness, demand analysis, and production cost can be used for many ground water management decisions that involve use values. Such methods offer defensible estimates of what are likely to be the major benefits of ground water services.**
- **The pervasiveness and magnitude of nonuse values for ground water is uncertain. Few and limited studies have been conducted, and little reliable evidence exists from which to draw conclusions about the importance of nonuse values for ground water. Additional research is needed to document the occurrence and size of nonuse values for ground water systems.**
- **What is most relevant for decision-making regarding ground water policies or management is knowledge of how the TEV of ground water will be affected by a decision. Pending documentation of large and pervasive nonuse values for ground water, it is likely that in many, but not all, circumstances, measurement of use values or extractive values alone will provide a substantial portion of the change in TEV relevant for decision-making.**
- **In some circumstances the TEV is likely to be largely composed of nonuse values. At the current time, pending documentation of large and pervasive nonuse values for ground water systems, this appears to be most likely when ground water has a strong connection to surface water and a decision will substantially alter these service flows. In these situations, focusing on use values alone could seriously mismeasure changes in TEV and will ill serve decision-making. Decision-makers should approach valuation with a careful regard for measurement of TEV using direct techniques that can incorporate nonuse values.**

LEGAL CONSIDERATIONS, VALUATION, AND GROUND WATER POLICY

The last two decades have brought changes in emphases in both technical and institutional issues related to ground water management. Due to society's misplaced perceptions of ground water's "pure" natural quality, there has been overemphasis over several decades on ground water quantity issues rather than quality issues. This has included the magnitude of water supplies being developed and associated costs. Quality considerations were mainly related to chlorides, nitrates, and the need for disinfection prior to human consumption. Since the mid-1970s increasing attention has been given to deteriorations in ground water quality. With ground water issues becoming more complex, the incorporation of economic valuation of ground water and other natural resources in decision-making takes on more urgency. This is especially true where a resource supports an ecosystem of national significance that not all citizens may be in contact with but still want protected (e.g., the Everglades or the Grand Canyon).

Sixteen federal laws relate directly or indirectly to ground water management. Key laws include the Clean Water Act (CWA), Safe Drinking Water Act (SDWA), Resource Conservation and Recovery Act (RCRA), Comprehensive Environmental Response, Compensation, and Liability Act (CERCLA, or Superfund), and Superfund Amendments and Reauthorization Act (SARA). The SDWA addresses the quality of public drinking water supplies and ground water protection. The CWA addresses pollution control, while RCRA relates to waste disposal sites and underground storage tanks. Soil and ground water remediation are the subjects of the Superfund laws (CERCLA and SARA). Numerous state and local laws also address ground water usage (quantity allocations) and quality via numerical standards or descriptive criteria. These multiple laws and regulatory agency overlaps can create conflicts regarding ground water usage, quality protection, and/or remediation responsibility.

Command-and-control approaches have historically dominated pollution control in environmental quality laws. More recently, market-based considerations, incentives for pollution prevention, and risk management have been advanced as additional components in environmental management, including the management of ground water. Many of these recent environmental management approaches include consideration of some economic issues, including program or project costs and benefits.

Water marketing (the buying and selling of water rights) has emerged as a valuable policy alternative for allowing water allocation laws to efficiently respond to all water use demands. Theory suggests that where price reflects the TEV, reliance on water marketing is a more efficient way to allocate scarce resources.

On a national level, regulatory impact assessment has been used to address some economic issues. For example, President Reagan initiated a formal balancing of the benefits of environmental protection and regulatory compliance costs through Executive Order 12291, which required EPA and other agencies to prepare benefit-cost analyses for any proposed regulations imposing public and private costs of at least $100 million annually. Presidents Bush and Clinton issued similar Executive Orders. Current congressional interests include expanded use of risk assessments coupled with economic evaluations for both programs and projects.

At times, specific legislative mandates or principles may take precedence over the consideration of economic valuation information or benefit-cost analyses. Most federal environmental, health, and safety programs contain program requirements that are unfunded mandates. Accurate information regarding ground water values would make unfunded mandate regulatory reviews better relative to evaluation of the economic and environmental trade-offs involved in ground water protection policies. Historical ground water allocation schemes and water rights laws are examples, as is the concern over human health effects and their immediate reduction in the near-term requirements of the Superfund laws.

These institutional considerations suggest several areas of governmental action:

- **Federal, state, and local agencies should give consideration to the TEV of ground water in their deliberations on new or amended legislation or regulations related to ground water management.**
- **States should consider the authorization and promotion of water marketing, including transfer of ground water rights when appropriate. Although a transition to a market that adequately captures the full value of the resource may be difficult, water markets provide flexibility in water use and more efficient allocation of water among uses. Water markets also provide real world prices of water for current use values, and their prices aid decision-makers in valuing ground water.**
- **States should be encouraged to develop clear and enforceable rights to ground water where such rights are either lacking or absent. A system of clear and enforceable extractive rights to ground water is a prerequisite to economically efficient use of that water. Without such rights, users lack the incentive to value ground water appropriately (consideration of the full TEV) either now or in the future.**
- **EPA and other pertinent agencies should plan and implement an integrated and comprehensive research effort on ground water valuation. Federal agencies should conduct research and develop case studies in ground water valuation that includes a range of environmental conditions and economic circumstances. In addition, governmental agencies should sponsor further research jointly with private institutions to develop valuation methods that quantify ecological services and values. The results of such research will assist states in managing and protecting their ground water resources and could help to demonstrate improvements that valuation can bring to decision-making.**

CASE STUDY OBSERVATIONS

Chapter 6 contains brief synopses of seven case studies in which ground water valuation has been or could be used to enhance problem analysis and the decision-making process. The case studies illustrate different themes associated with the integration of hydrogeological, ground water usage and economic valuation information in real-world decision contexts. The Treasure Valley, Oregon, case illustrates the role of ground water in ecological services and how valuation can be incorporated in the allocation of scarce water supplies. The Laurel Ridge, Pennsylvania, study focuses on institutional fragmentation and the need for a watershed approach in ground water valuation and management. A study of Albuquerque, New Mexico examines the importance of hydrological information and the interaction of ground and surface waters in developing a long-term sustainable ground water policy. The Arvin-Edson, California, study illustrates the buffer value of ground water relative to extractive services in an area subject to surface water drought conditions. The Orange County, California, case study emphasizes the value of artificial recharge as a means of averting the loss of a ground water supply due to sea water intrusion. A Woburn, Massachusetts, example describes the use of benefit-cost analysis to integrate valuation information in a Superfund remediation dilemma. Finally a water supply study for Tucson, Arizona illustrates planning considerations associated with the valuation framework in Chapter 3, the methods illustrated in Chapter 4, and the importance of substitute water supplies.

These case studies offer several lessons, with most of them supportive of earlier conclusions. Among other things, they show that TEV provides a useful context for the qualitative recognition and/or quantitative valuation of ground water services. At the same time, each study is unique, thus limiting opportunities for subsequent benefits transfer analysis; and highlighting the technical, economic, institutional, and political uncertainties characterize the current state-of-the-art of ground water valuation.

1

Introduction and Background

THE GROUND WATER VALUATION DILEMMA IN BRIEF

Typically, water in the United States has not been traded in markets. Because of this, there are no market-generated prices or meaningful estimates of the value that markets would assign to water, if in fact water were a traded good. This undetermined value for water is most apparent in the case of ground water. Whatever might have been the historic circumstances, there is no basis today for our practice of judging the value of ground water to be negligible. All scarce resources, commodities, and services have value. Ground water is often a scarce resource, whether judged by the direct use people make of it (for example, as drinking water) or by its less obvious ecological functions, such as wetlands maintenance and its contribution to stream flow; or the prevention of land subsidence.

The longer we ignore or distort ground water's value, the more overused, degraded, and misallocated the resource becomes. Without price signals or other indicators of value to help guide policy, we tend to devote too little attention and funding for resource management and protection of ground water.

Goods or services that are generally not "owned" in the same sense as other property are often not traded in well-functioning markets. Without such markets, determining the value of these goods and services becomes more complicated, and analytical efforts must be made to estimate values. In the case of ground water, that estimate must account for both the cost of pumping and delivery and the inherent value of the resource, reflecting its multiple services, the "goods" it provides and the "costs or hardships" it protects against. In many states and localities, however, the charge to the user is confined to out-of-pocket costs such as energy for pumping and amortization of investments in well construction and costs of treatment and distribution systems. These are necessary components of the value of ground water. But undervaluation of the resource is inevitable, principally because there is no widely accepted means of recording users' or society's valuation of those broader use and nonuse attributes.

Improved ground water valuation techniques and estimates could assist water resource management and policy-making in many important ways. For example, an improved ability to weigh alternative water sources or protection strategies should lead to better allocation of scarce Superfund dollars. There is general agreement that water resource decision-making has focused mainly on an evaluation of alternative projects primarily by the costs of these projects. However, improved techniques would facilitate the decision-making for cleanup and protection based on a better standard, one that compares values and benefits of different ground water sites.

The valuation principles described in this report can be a critical input to but are distinct from cost-benefit analysis. That is because the estimation of *costs* (for example, the infrastructural investment requirements of a municipal water system) is of secondary concern here. To be sure, certain nonmarket values at risk on the cost side, such as subsidence, increased salinity from excessive ground water mining, wetland degradation, and destruction of riparian habitat are relevant. The principal emphasis here is on methods that value the benefits of ground water.

Ground water valuation concepts and challenges discussed in the following chapters cut across numerous valuation dilemmas in the natural resources-environmental arena. An example of such similarity and overlap is the problem of assigning values to *surface* water. Of course, surface water and its management provide some unique services (e.g., navigation, power, and flood control) not applicable to ground water, but many services are common to both surface and ground water (household use, irrigation, and joint ecological benefits, such as wetlands maintenance). Moreover, ground and surface water are hydrologically linked, so that the contamination of one body can migrate to the other. There is no way to divide up benefits neatly and analyze value simply.

CONTEXT FOR GROUND WATER VALUATION

Trends in Ground Water Use and Protection

Tables 1.1 and 1.2 and Figure 1.1 provide quantitative highlights of trends in U.S. water use. The growth of withdrawals of ground or surface water from 1950 to 1990, occurred largely during the first 25 years of that time span, substantially

TABLE 1.1 Withdrawals of Water, by Type and Category of Use, 1990

		Percent of	
	Million gals. per day	Total U.S.	Ground or Surface
U.S. Total	408,000	100.0	
Ground water, total	80,620	19.8	100.0
Public supply	15,100	3.7	18.7
Domestic	3,260	0.8	4.0
Commercial	908	0.2	1.1
Irrigation	51,000	12.5	63.3
Livestock	2,690	0.7	3.3
Industrial	3,960	1.0	4.9
Mining	3,230	0.8	4.0
Thermoelectric	525	0.1	0.7
Surface water, total	327,000	80.1	100.0
Public supply	23,500	5.8	7.2
Domestic	132	0.0	0.0
Commercial	1,480	0.4	0.5
Irrigation	85,500	21.0	26.1
Livestock	1,800	0.4	0.6
Industrial	18,600	4.6	5.7
Mining	1,718	0.4	0.5
Thermoelectric	194,500	47.7	59.5

SOURCE: Compiled from Solley et al., 1993. Because of rounding, individual items may not add precisely to totals.

exceeding U.S. population growth in that period. Since 1975 water use has remained essentially flat. The U. S. Geological Survey (USGS) singles out three factors to account for that level trend. First, higher energy prices and declines in farm commodity prices in the 1980s reduced the demand for irrigation water and spurred the introduction of more efficient pumping technologies. In addition, pollution control regulations encouraged recycling and reduced discharge of pollutants, thereby decreasing water requirements in the industrial sector. And more generally, the public became increasingly concerned about conservation (Solley et al., 1993). No doubt the slowdown in development of new hydroelectric capacity in the United States contributed as well. However, the USGS does not identify water pricing as a factor in the deceleration of water use, though higher energy prices would have constituted an indirect disincentive to consumption.

Ground water is the predominant source of water supply for rural areas in the United States, primarily for agriculture and domestic use. In 1985 ground water provided drinking water for more than half the U.S. population and 97% of the rural population (Moody, 1990). As Table 1.1 indicates, agriculture (irrigation

TABLE 1.2 Trends of Estimated Water Use in the U.S., 1950-90

	1950	1955	1960	1965	1970	1975	1980	1985	1990	Average annual % rate of change 1950-75	Average annual % rate of change 1975-90
Population (mill.)	150.7	164.0	179.3	193.8	205.9	216.4	229.6	242.4	252.3	1.5	1.0
Withdrawals (bill. gals. per day)	180	240	270	310	370	420	440	399	408	3.4	–0.2
Ground	34	48	50	60	69	83	84	74	81	3.6	–0.2
Surface	150	198	221	253	303	329	361	325	327	3.2	–0.0

SOURCE: Compiled from Solley et al. (1993). Because total withdrawals were rounded off, the ground and surface numbers do not add precisely to totals.

and livestock) uses approximately two-thirds of the total ground water withdrawn in the United States, with public supply (including domestic withdrawals) accounting for nearly a quarter of the total.

In the late 1970s and early 1980s, the U.S. Environmental Protection Agency (EPA) began to direct attention toward ground water pollution studies, emphasizing the identification and evaluation of pollution sources and source categories and subsurface transport and fate processes for both inorganic and synthetic organic chemicals. Also in the early 1980s, the inception of the Superfund program brought attention to the need to clean up contaminated soil and ground water and led to major remediation programs by EPA and the Departments of Defense and Energy. In 1984 EPA adopted a ground water protection strategy that focused on land use planning, engineering control measures, and management practices that could be used to prevent ground water contamination and thus protect ground water quality.

The Safe Water Drinking Act of 1986 included a wellhead protection program to further encourage such pollution prevention efforts by state and local governments. EPA has continued to promulgate policies and related guidance to stress the importance of protecting renewable ground water resources from contamination and thus minimize the need for remediation efforts (U.S. EPA, 1991).

Ground Water Valuation Terminology

The inherently interdisciplinary nature of the ground water valuation problem becomes obvious in the confusion about terminology used to describe it. There is no commonly used ground water valuation terminology and no one set that is obviously superior. Two different sets of valuation terminology are displayed in Table 1.3. The first is based upon the physical state of the ground water from which value is derived. The primary distinction is between extractive values, which occur as a result of the extraction of ground water and subsequent consumptive use, and *in situ* values, which occur as a consequence of leaving the water in the aquifer. Extractive values include municipal, agricultural, and industrial uses of water, uses that nearly always include a sizable component of consumptive use. *In situ* values are derived from the services provided by leaving water in the aquifer and typically do not involve consumptive transformation of the water. *In situ* values include ecological values, buffer values, values associated with the avoidance of subsidence, recreational values, existence values, and bequest values.

The second set of terminology comes from the economic literature or the valuation of ground water resources which classifies ground water values in terms of use values and nonuse values. This distinction acknowledges that use values are associated with both consumptive and nonconsumptive uses of water, including ecological uses, buffering, subsidence avoidance, and recreation. By contrast, nonuse values, including existence and bequest value, may occur when

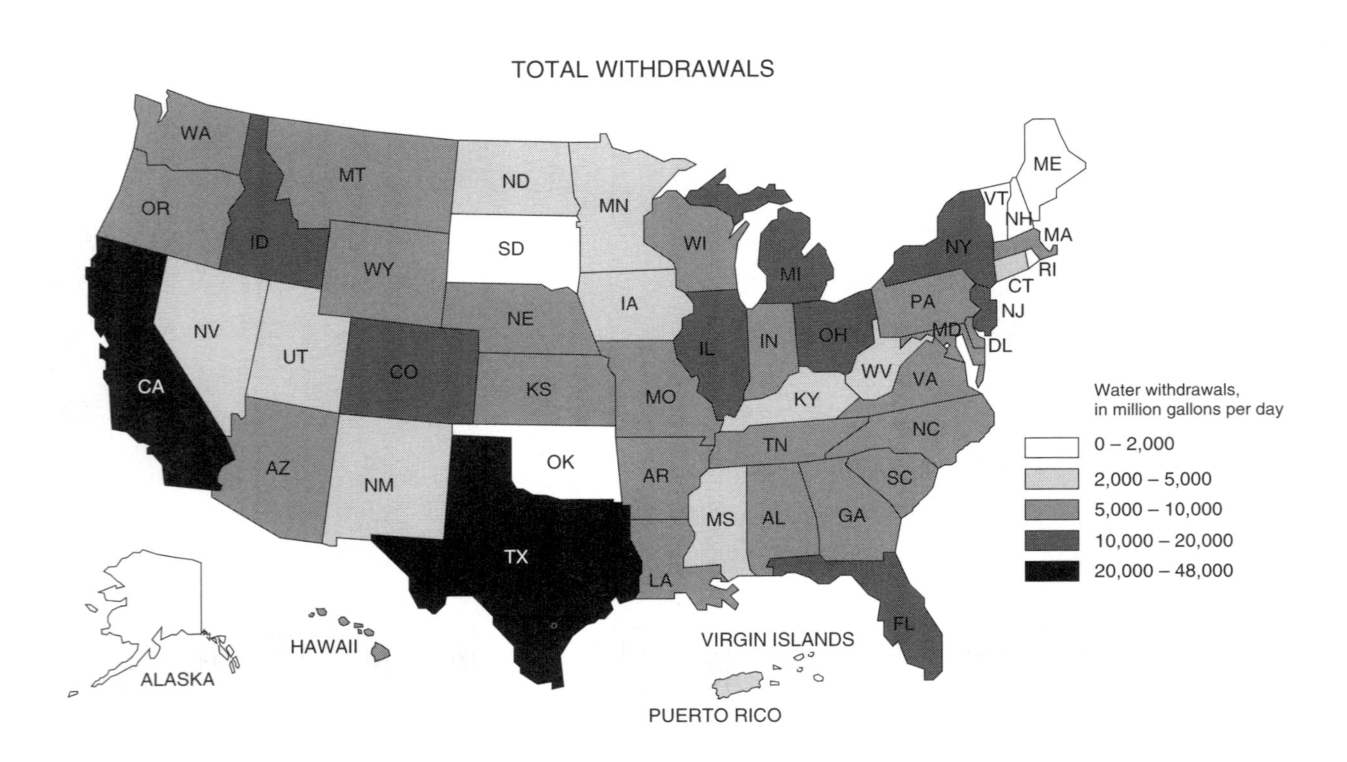
TOTAL WITHDRAWALS
WA
OR
CA
NV
ID
MT
WY
UT
AZ
CO
NM
ND
SD
NE
KS
OK
TX
MN
IA
MO
AR
LA
WI
IL
MI
IN
OH
KY
TN
MS
AL
GA
FL
SC
NC
VA
WV
PA
NY
MD
DL
NJ
CT
RI
MA
VT
NH
ME
ALASKA
HAWAII
VIRGIN ISLANDS
PUERTO RICO
Water withdrawals, in million gallons per day
0 – 2,000
2,000 – 5,000
5,000 – 10,000
10,000 – 20,000
20,000 – 48,000

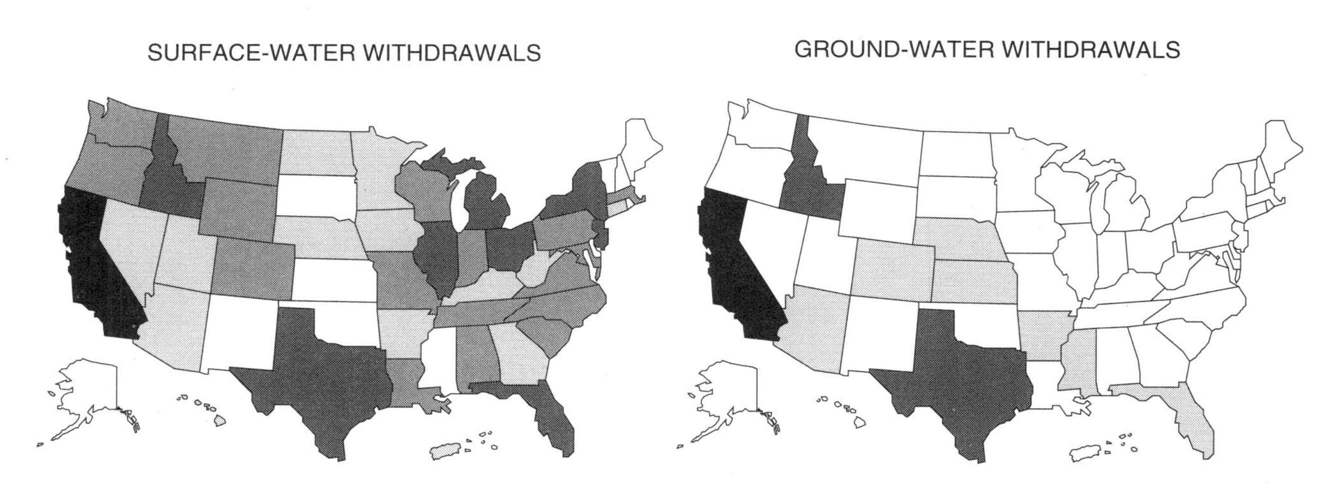

FIGURE 1.1 Total water withdrawals by source and state, 1990. SOURCE: Solley et al., 1993

TABLE 1.3 Taxonomy of Ground Water Valuation Terminology

Physical State Terminology	Economic Terminology	Accounting Terminology: Stocks	Accounting Terminology: Flows
A. Extractive values			
1. Municipal use values	Use Values		*
2. Industrial use values	Use Values		*
3. Agricultural use values	Use Values		*
4. Other extractive use values	Use Values		*
B. *In situ* values			
1. Ecological values	Use Values	*	*
2. Buffer values	Use Values	*	*
3. Subsidence avoidance values	Use Values	*	*
4. Recreational values	Use Values		*
5. Sea water intrusion values	Use Values	*	*
6. Existence values	Nonuse Values		*
7. Bequest values	Nonuse Values		*

the ground water is not devoted to any use. The relationship between these two different sets of taxonomies is also depicted in Table 1.3. The terminology used in the remainder of this report follows these two taxonomies and the relationship between stocks and flows of ground water. It is important to recognize that it is sometimes difficult to draw a distinct line between use and nonuse values. For example, in southern California ground water in the aquifer has a value in protecting against sea water intrusion. Sea water intrusion can affect both the use values of ground water, by increasing the cost of drinking water supplies, and the nonuse values, through contamination of the aquifer even if it was never to be used as source of water for human consumption.

Unless care is taken in definition and use, both sets of terminology may mask or confuse the important distinction between values which are associated with a flow or stream of goods and services and values which are associated with stocks or assets which create those streams. Flows or streams of value, such as use values which come from extraction, recur over time and contrast with stock values which are the value of an asset (or liability) which yields flows of value over time. Flow values and stock values are linked because a stream of values (costs or benefits) can be converted into an asset value by calculating the present discounted value of the flow. Failure to distinguish between the value of flows and the value of a stock or asset may result in double counting or other errors. The last two columns of Table 1.3 indicate whether the various categories of physical state or economic values are commonly treated as flows or stocks or as both. In instances where values are commonly expressed as either stocks or flows, it is important to specify whether the value is a flow value or a stock (asset value).

Services Provided by Ground Water

Tables 1.4 and 1.5 encapsulate some of the major services ground water provides that give rise to economic value. (Detailed discussion of the approaches used to determine quantitative estimates of value is in Chapter 4.) Although many people already appreciate or may have intuitively accepted the nature of these services and the importance of assigning value to them, they have not engaged in widespread action to more effectively conserve, protect, and allocate these resources. Part of the problem no doubt arises from the technically demanding nature of the problem, for example, the complex behavior and properties of aquifers. In particular it is challenging to evaluate from an economic perspective the ecological services rendered by ground water since such services are not traded in markets and are viewed in a highly subjective way.

Table 1.6 (also in Chapter 4 as Table 4.5) presents an overview of the alternative valuation methods for addressing selected ground water functions/ services. These economic valuation methods and existing applications are discussed in more detail in Chapter 4.

Ground water problems are receiving more attention for a number of reasons. Increased withdrawals are causing problems such as subsidence, salt water intrusion, and destruction of wildlife habitat. Public water supply systems dependent on ground water can be found in every state (Solley et al., 1993). Also, the importance of ground water as a buffer, or emergency supply, is beginning to be more widely recognized. This value was illustrated in California during the drought in the early 1990s, when demands for surface water far outstripped the available supplies. Agricultural and municipal ground water use increased dramatically, causing concerns about whether ground water protection regulations were adequate.

The importance of ground water has also changed in the context of conjunctive use. Recharge of surface water in Florida and use of effluent to replenish ground water is now common in southern California. In this context the ground water aquifer becomes an actively managed storage facility, with ground water supplies replenished by flood flows, imported surface water, and treated effluent. Water is cycled through the aquifer materials on a relatively short term basis and provides a buffer against shortages of surface water.

The environmental values associated with ground water are also becoming more widely recognized. Just as the ecosystem concept is gaining more recognition in habitat management to protect animal species, the role of ground water in the support of surface water supplies, wetlands, and riparian habitat is more clearly understood.

Management/Regulatory Decisions Related to Valuation

Most decisions regarding ground water development, use, or protection are

TABLE 1.4 Potential Service Flows and Effects of Those Services for Ground Water Stored in an Aquifer

Service Provided	Effect on Value
Potable water for residential use	Change in Availability of Potable Water Change in Human Health or Health Risks
Landscape and turf irrigation	Change in Cost of Maintaining Public or Private Property
Agricultural crop irrigation	Change in Value of Crops or Production Costs Change in Human Health or Health Risks
Livestock watering	Change in Value of Livestock Products or Production Costs Change in Human Health or Health Risks
Food product processing	Change in Value of Food Products or Production Costs Change in Human Health or Health Risks
Other manufacturing processes	Change in Value of Manufactured Goods or Production Costs
Heated water for geothermal power plants	Change in Cost of Electricity Generation
Cooling water for other power plants	Change in Cost of Electricity Generation
Prevention of land subsidence	Change in Cost of Maintaining Public or Private Property
Erosion and flood control through absorption of surface water runoff	Change in Cost of Maintaining Public or Private Property
Medium for wastes and other by-products of human ecomic activity	Change in Human Health or Health Risks Attributable to Change in Ground Water Quality Change in Animal Health or Health Risks Attributable to Change in Ground Water Quality Change in Economic Output Attributable to Use of Ground Water Resources as "Sink" for Wastes
Improved water quality through support of living organisms	Change in Human Health or Health Risks Attributable to Change in Ground Water Quality Change in Animal Health or Health Risks Attributable to Change in Ground Water Quality Change in Economic Output or Production Costs Attributable to Use of Ground Water Resources as "Sink" for Wastes
Nonuse services (e.g., existence or bequest motivations)	Change in Personal Utility

SOURCE: Modified from Boyle and Bergstrom, 1994.

TABLE 1.5 Potential Service Flows and Effects of Those Services for Surface Water and Wetland Surfaces Attributable to Ground Water Reserves

Service Provided	Effect on Value
Surface water supplies for drinking water	Change in Availability of Potable Water Change in Human Health or Health Risks
Surface water supplies for landscape and turf irrigation	Change in Cost of Maintaining Public or Private Property
Surface water supplies for agricultural crop irrigation	Change in Value of Crops or Production Costs Change in Human Health or Health Risks
Surface water supplies for watering livestock	Change in Value of Livestock Products or Production Costs Change in Human Health or Health Risks
Surface water supplies of food product processing	Change in Value of Food Products or Production Costs Change in Human Health or Health Risks
Surface water supplies for manufacturing processes	Change in Value of Manufactured Goods or Production Costs
Surface water supplies for power plants	Change in Cost of Electricity Generation
Erosion flood and storm protection	Change in Cost of Maintaining Public or Private Property Changes in Human Health or Health Risks through Personal Injury Protection
Transport and treatment of wastes and other by-products of human economic activity through surface water supplies	Change in Human Health or Health Risks Attributable to Change in Surface Water Quality Change in Animal Health or Health Risks Attributable to Change in Surface Water Quality Change in Economic Output or Production Costs Attributable to Use of Surface Water Resources for Disposing of Wastes
Recreational swimming, boating, fishing, hunting, trapping, and plant gathering	Change in Quality or Quantity of Recreational Activities Change in Human Health or Health Risks
Commercial fishing, hunting, trapping, and plant gathering supported by ground water discharges	Change in Value of Commercial Harvest or Costs Change in Human Health or Health Risks
On-site observation or study of fish, wildlife, and plants purposes supported by ground water discharges for leisure, educational, or scientific purposes	Change in Quantity or Quality of On-Site Observation or Study Activities

(continued)

TABLE 1.5 (continued)

Service Provided	Effect on Value
Indirect, off-site fish, wildlife, and plant uses (e.g., viewing wildlife photos)	Change in Quality or Quantity of Indirect Off-Site Activities
Improved water quality resulting from living organisms related to ground water discharges	Change in Human Health or Health Risks Attributable to Change in Air Quality Change in Animal Health or Health Risks Attributable to Change in Air Quality Change in Value of Economic Output or Production Costs Attributable to Change in Air Quality
Regulation of climate through support of plants	Change in Human Health or Health Risks Attributable to Change in Climate Change in Animal Health or Health Risks Attributable to Change in Climate Change in Value of Economic Output or Production Costs Attributable to Change in Climate
Provision of nonuse services (e.g., existence services) associated with surface water bodies or wetlands environments or ecosystems supported by ground water discharges	Change in Personal Utility or Satisfaction

SOURCE: Modified from Boyle and Bergstrom, 1994.

made with inadequate attention to the value of ground water as a source of consumptive use and for the *in situ* services it provides. For example, although many states require permits to drill a new high-capacity well, they tend to grant such permits on a routine basis, neglecting the broad range of values at stake. Management decisions have traditionally been made by comparing the direct financial costs of various alternatives, without taking into account the impacts on the full set of values of ground water. This tendency to consider only financial costs limits the usefulness of the underlying calculations for cost-benefit analysis. As a result, ground water (which should be a renewable resource) tends to be rationally managed only where problems of depletion or pollution are apparent or have become critical. In most areas of the country, the ground water management policy may be aptly characterized as out of sight, out of mind.

Superfund: Prevention vs. Remediation

In some cases ground water quality degradation may be irreversible. In such situations it becomes especially important that the resource is properly valued. If

TABLE 1.6 A General Matrix of Ground Water Functions/Services and Applicable Valuation Methods

Ground Water Function/Service Flow	Applicable Valuation Method
A. Extractive values	
1. Municipal use (drinking water)	
a) Human health - morbidity	Cost of illness Averting behavior Contingent valuation Contingent ranking/behavior
b) Human health - mortality	Averting behavior Contingent valuation Contingent ranking/behavior
2. Agricultural water use	Derived demand/production cost
3. Industrial water use	Derived demand/production cost
B. *In situ* values	
1. Ecological values	Production cost techniques Contingent valuation Contingent ranking/behavior
2. Buffer value	Dynamic optimization Contingent valuation Contingent ranking/behavior
3. Subsidence avoidance	Production cost Hedonic pricing model Contingent valuation Contingent ranking/behavior
4. Recreation	Travel cost method Contingent valuation Contingent ranking/behavior
5. Existence value	Contingent valuation Contingent ranking/behavior
6. Bequest value	Contingent valuation Contingent ranking/behavior

SOURCE: Adapted from Freeman, 1993. (Reprinted with permission from Resources for the Future, 1993. Copyright 1993 by Resources for the Future.)

ground water of suitable quality becomes increasingly scarce owing to pollution and if substitute sources are unavailable, then the resource's value may rise abruptly. Conversely, a contaminated aquifer that poses little threat to its ambient surroundings and possesses few prized attributes not available from substitutes would rate no such premium.

At both the federal and state levels, there have been prodigious efforts in recent years to remediate the subsurface environment for purposes of ground water quality protection and restoration. Current estimates of the total costs of remediation, through Superfund and analogous programs in the Departments of Energy and Defense as well as state and local efforts, amount to hundreds of billions of dollars. Consideration of this enormous expense has led the regulated community and some decision-makers to question the benefit-cost balance of mandated subsurface remediation programs.

The Committee on Ground Water Cleanup Alternatives of the National Research Council (NRC) has recently reviewed the technical means to restore ground water quality (NRC, 1994). The committee found that there is no panacea for treating ground water contaminated by hazardous wastes. Especially in cases with heterogeneous hydrogeologic conditions and complex chemical behavior, it may prove infeasible to restore ground water to its "pristine" state. In such instances it may be necessary to revert to strategies that aim to contain, or isolate, the contamination to the extent possible, thus alleviating the endangerment of surrounding ground water supplies. However, even the less ambitious objective of containment implies the obligation indefinitely to monitor the quality of the adjacent threatened ground water, as well as to remove the maximum feasible mass of contaminants in order to minimize the consequences of possible failure of the containment measures. For now, the debate continues as to whether a comprehensively implemented containment strategy will prove less expensive in the long term than the current policy of complete cleanup.

Valuation, including consideration of alternative uses of an affected site, and the costs of alternative sources of water, would not only be a useful tool to guide decisions on whether to pursue containment or remediation but is also worthwhile for clarifying various tradeoffs to contamination prevention action. Increasing awareness of the need to prevent contamination of ground water supplies, on top of mounting costs of remediation, point to the importance of coordinated and comprehensive land and water use decision-making. Only within such a broad framework will it be possible to inject ground water valuation into strategies for containment, remediation, or alternatives for safeguarding the welfare of the community.

Management Issues

Water managers make decisions within a particular sociopolitical and technical context. They are constrained by technical considerations such as capacity of various conveyance facilities, recharge capability of an aquifer, physical availability of surface water supplies, and environmental or resource impacts of supply development. They are also limited by the institutional environment in which they operate, including federal, state, and local regulations and court-decreed rights and uses of ground water, and legislated or adjudicated mandates are not always in accord with economically optimal outcomes. Financial constraints can greatly aggravate the political landscape; the impact that a particular course of action has on local water rates and taxes is frequently the controlling factor in a water management decision.

The public has become progressively more involved in water-related decision-making in the past few decades. Because of the pervasive importance of water availability in virtually all types of activity (in households, commercial development, industry, and agriculture) water issues are commonly linked with concerns about economic growth. Water management decisions are often infused with local or regional politics and burdened with a heavy overlay of social values. As a result, the degree of autonomy of local and regional water providers varies greatly among and within states.

History of Economic Valuation of Natural Resources

The principles for valuing natural assets such as energy and mineral deposits, forests, and aquifers were set forth more than 60 years ago. This research established a relationship between the value of the asset and the present value of the services it provides. Some of the earliest attempts to value nonmarketed goods and services focused on environmental and natural resource assets. One of the first such efforts involved the development of value measures for water resources used in irrigated agriculture in the western United States. Linear programming models and other techniques were used to estimate the value of both surface and ground water by examining how the profitability of farm enterprises changed as water became more or less available. These techniques worked well in assigning economic value to water use in agriculture since water is an input to the production processes of firms whose products are sold in reasonably well functioning markets. These early methods, which highlighted the valuation of nonmarketed *inputs*, were not well adapted for measuring the value of nonmarketed *outputs* that are consumed directly.

Principles for measuring the consumptive value of water for household use were set forth a century ago. The idea of using a demand curve to measure the value of a good as the area under the demand curve (consumer surplus) was articulated in the late 19th century, and applied methods for doing so have been developed ever since. Hewitt and Hanemann (1995) provide a sophisticated example of an application to urban water.

Two techniques were developed specifically for the estimation of nonmarketed outputs: the travel cost method (TCM) and the contingent valuation method (CVM). The first, created to value visits to national parks, is an example of an indirect methodology to infer values of nonmarketed goods and services by examining ancillary evidence such as expenditures on travel. Refinements in the TCM and the development of other indirect techniques have enhanced the ability of economists to value a wide range of natural resource and environmental services, including improvements in air and water quality. These indirect techniques, however, are sometimes based on questionable assumptions and often require the resolution of difficult and complex problems in statistical estimation.

Using an earlier suggestion by Ciriacy-Wantrup; Davis in 1963 undertook the first application of stated-preference approaches to valuing a natural asset (Ciriacy-Wantrup, 1952; Davis, 1963). These CVM techniques rely on carefully structured interviews with consumers and potential consumers to elicit measures of economic value (see Appendix B). Such direct techniques have proved useful in measuring the value of a wide range of goods and services not traded in a market. In particular CVM techniques have been used to estimate nonuse values. Nevertheless, direct valuation techniques, like their indirect counterparts, are subject to both conceptual and practical difficulties.

Although nonmarket valuation techniques have been helpful in valuing individual environmental commodities, policy and regulatory attention has increasingly focused on the management of ecosystems. The need to value complex hydrologic or ecological functions and the associated range of service flows raises a number of issues in nonmarket valuation. Part of the difficulty in valuing ecosystem services is that ecologists cannot define and measure unambiguously the performance of ecosystems and boundaries of successional trajectories. Other problems arise from the inability of economists to measure the consequences of complex phenomena over the long run. Further problems grow out of differences in disciplinary perspectives, which complicate the interdisciplinary task of integrating the physical relationships required for bioeconomic assessments.

THE ROLE OF THE NRC

The Environmental Protection Agency requested that the NRC appoint a committee to study approaches to assessing the future economic value of ground water and the economic impact of the contamination or depletion of these resources. In 1994 the NRC appointed a committee to conduct this study under the auspices of the NRC's Water Science and Technology Board. The committee was charged to:

(1) review and critique various approaches for estimating the future value of uncontaminated ground water in both practice and theory (addressed in Chapters 2, 3, and 4);

(2) identify areas in which existing approaches require further development and promising new approaches might be developed (addressed in Chapters 3 and 4);

(3) delineate the circumstances under which various approaches would be preferred in deciding long-term resource use and management (addressed in Chapters 4 and 6);

(4) outline legislative and policy considerations in connection with the use and implementation of recommended approaches and related research needs (addressed in Chapter 5); and

(5) illustrate, through real or hypothetical case examples, how recommended procedures would be applied in practice for representative applications (addressed in Chapter 6).

The committee's report is organized into six chapters. Chapter 2 addresses ground water hydrology, ecology, and economic concepts relevant to valuation studies. Chapter 3 highlights the relationship between time, institutional, and hydrologic constraints and ground water services; it goes on to explain extractive and *in situ* services. Also included is a conceptual framework for calculating economic value based on services, modified from Boyle and Bergstrom (1994). The central concept of total economic value and the role of time/discounting and uncertainty round out Chapter 3.

A critique of valuation methods, for example, the contingent valuation method, the travel cost method, and the hedonic pricing method, as applied in ground water-related studies is the focus of Chapter 4. Advantages and limitations of such methods are described along with their application in delineating use and nonuse values for ground water resources. The available evidence from existing ground water valuation studies is compared with the possible range of extractive and *in situ* values identified in earlier chapters. Chapter 5 explores how various institutional issues such as ground water law and allocation methods can both affect and be improved by valuation study results. The last chapter contains brief synopses of seven case studies in which ground water valuation has been or could be used to enhance problem analysis and the decision-making process.

This report blends both resource depletion (ground water mining) issues with quality deterioration issues as they relate to valuation. Further, there are relationships between depletion and quality which need to be recognized. Finally, the reader should be aware that these issues and relationships are, of necessity, intertwined throughout the report.

REFERENCES

Boyle, K. J., and J. C. Bergstrom. 1994. A Framework for Measuring the Economic Benefits of Ground Water. Department of Agricultural and Resource Economics Staff Paper. Orono: University of Maine.

Ciriacy-Wantrup, S. V. 1952. Resource Conservation. Berkeley: University of California.

Davis, R. 1963. The Value of Outdoor Recreation: An Economic Study of the Maui Woods. Ph.D. dissertation, Harvard University.

Hewitt, J. A., and W. M. Hanemann. 1995. A discrete/continuous choice approach to residential water demand under block-rate pricing. Land Economics 71(2):173-192.

Freeman, A. M., III. 1993. The Measurement of Environmental and Resource Values: Theory and Methods. Washington, D.C.: Resources for the Future Press.

Moody, D. W. 1990. Ground water contamination in the United States. Journal of Soil and Water Conservation 45(2):170-179.

National Research Council. 1994. Alternatives for Ground Water Cleanup. Washington, D.C.: National Academy Press.

Solley, W. B., R. R. Pierce, and H. A. Perlman. 1993. Estimated Use of Water in the United States in 1990. U.S. Geological Survey Circular 1081. Washington, D.C.: U.S. GPO.

U.S. Environmental Protection Agency. 1991. Preliminary Risk Assessment for Bacteria in Municipal Sewage Sludge Applied to Land. EPA/600/6-91/006. Cincinnati: U.S. Environmental Protection Agency.

2

Ground Water Resources: Hydrology, Ecology, and Economics

The use value of ground water depends fundamentally on the costs of producing or obtaining the water and its value in the uses to which it is ultimately put. The costs of producing ground water typically include the costs of extraction and delivery as well as the opportunity cost of using the water right away rather than leaving it in storage for later use. The value in alternative uses can be expressed by the willingness of users to pay for the water. Willingness to pay depends in turn upon a number of factors, including the quality of the water. The quality of ground water should be thought of in terms of its acceptability for certain uses. Thus the quality of a given source of ground water may not be acceptable for potable uses but may be sufficient for a wide variety of nonpotable uses. Because extraction and delivery costs are related to the quantity of ground water, the real question is, What is the availability of ground water that possesses some desired quality? Ground water quality and the costs of extraction depend on the geologic and hydrologic characteristics of a given aquifer as well as the economic circumstances that characterize the particular uses to which ground water is devoted. Both the current and future values of ground water, then, are determined jointly by the interaction of geologic/hydrologic factors and economic factors.

HYDROLOGICAL CONCEPTS

Ground water is usually found in subsurface formations known as aquifers, which may be a significant hydrological component of watersheds and basins. Basins and watersheds are similar in that all of the collected water within them drains through a single exit point. Basins differ from watersheds only in the perception of their size, with basins being much larger than watersheds and typically composed of many watersheds. In the United States, "basin" is often used to mean a large riverine drainage system. Within a watershed or basin, water moves both on and below the surface. Aquifers are generally bounded by subsurface divides similar to surface features that separate watersheds. Often the boundaries of basins are not as obvious as those of watersheds, and aquifers may underlie and be common to several surface watersheds. Geologic strata that are tilted counter to the topography can conduct water in the opposite direction from topographic surface slopes. Large, confined aquifers may underlie smaller, unconfined zones that conform more closely to the surface topography. Because aquifers may be connected, the availability and quality of the ground water within them may be regional issues, defined by both surface and subsurface topography. The three-dimensional nature of aquifers is not generally well understood and is rarely considered in modeling for management applications. The condition and characteristics of a given aquifer are determined by the hydrologic cycle and by anthropogenic modifications in the hydrologic cycle.

The Hydrologic Cycle

The hydrologic cycle can be usefully depicted on both global and basinwide scales. In the global hydrologic cycle, water can be transferred from one location to another and transformed among the solid, liquid, and gaseous phases, but the total amount of water remains the same. From a basinwide perspective, the fact that water can be transferred from one basin to another means that specific basins can experience gains and losses in the total amount of water. This is an important concept in that the quantity of water in a basin can be depleted, whereas the total amount of water remains the same as it cycles among the various basins. The hydrologic cycle is depicted from a basin perspective in Figure 2.1. Precipitation is the pathway by which water enters the basin. Evaporation and transpiration, along with stream flows, are the principal pathways by which it leaves. Runoff, which is overland flow, can be augmented by interflow, which operates below the surface but above the water table, and by base flow, which refers to the discharge to streams from the saturated portion of the system. Infiltration of water into the subsurface is the ultimate source of both interflow and recharge to the ground water. Ground water recharge, defined as the portion of infiltration water that reaches the ground water, represents the replenishment of ground water supply.

Ground Water Balance: Recharge and Depletion

The quantity of water stored in an aquifer can be characterized over time by accounting for inflows and outflows according to the following expression:

$$\textit{Change in storage} = \textit{recharge} - \textit{depletion}$$

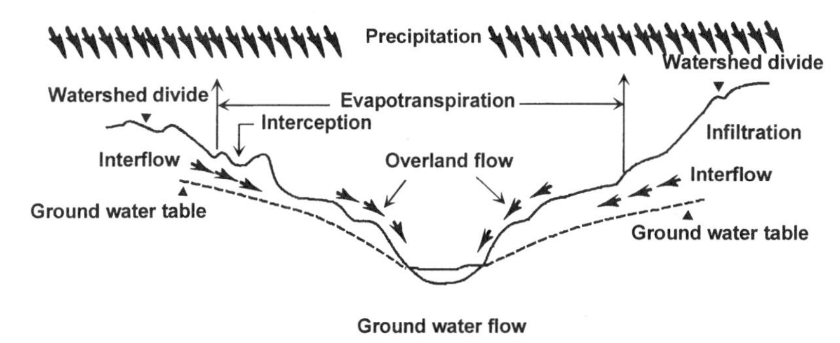

FIGURE 2.1 The hydrologic cycle as applied to basins.

Recharge occurs whenever precipitation or surface water infiltrates downward through the soil to the water table. Recharge can also result from subsurface lateral flows that reach the aquifer. Recharge may occur naturally, and natural recharge can be augmented by artificial recharges (as outlined in a recent study, National Research Council, 1994a). Surface water is usually viewed as a renewable resource, since it derives from rainfall and snowmelt, which recur periodically. Natural ground water supplies may be either renewable or nonrenewable, depending upon whether recharge occurs at rates similar to those of withdrawal.

The rate of recharge may be influenced to a large extent by whether the aquifer is confined or unconfined. Aquifers may have upper and lower boundaries, termed "confining layers." These boundaries normally comprise layers of unconsolidated material or rock that have a much lower permeability than the material lying immediately above or below. Confined aquifers have a confining layer both above and below, while an unconfined aquifer has no confining layer on top. Since unconfined aquifers tend to be found uppermost in a ground water system, they are frequently called surficial aquifers. Unconfined aquifers are the first to receive water infiltrating from the surface. This means that the depth to water or the water table frequently fluctuates in such aquifers. It also means that such aquifers tend to contain higher concentrations of dissolved materials of anthropogenic origin than do lowerlying, confined aquifers. Indeed, water contained in many shallow, unconfined aquifers is often not used for drinking because of contamination.

Confined aquifers are protected to some degree by the presence of a confining, low-permeability zone between the surface (and the source of recharge water) and the ground water itself. While an unconfined aquifer is characterized by a water table or the depth to ground water, a confined aquifer is characterized by a piezometric, or potentiometric, surface, which results because the height of the upper surface of the aquifer is constrained by the confining layer. The potentiometric surface represents the height of rise of the water due to hydrostatic pressure when the constraint of the confining layer is removed, as illustrated in Figure 2.2.

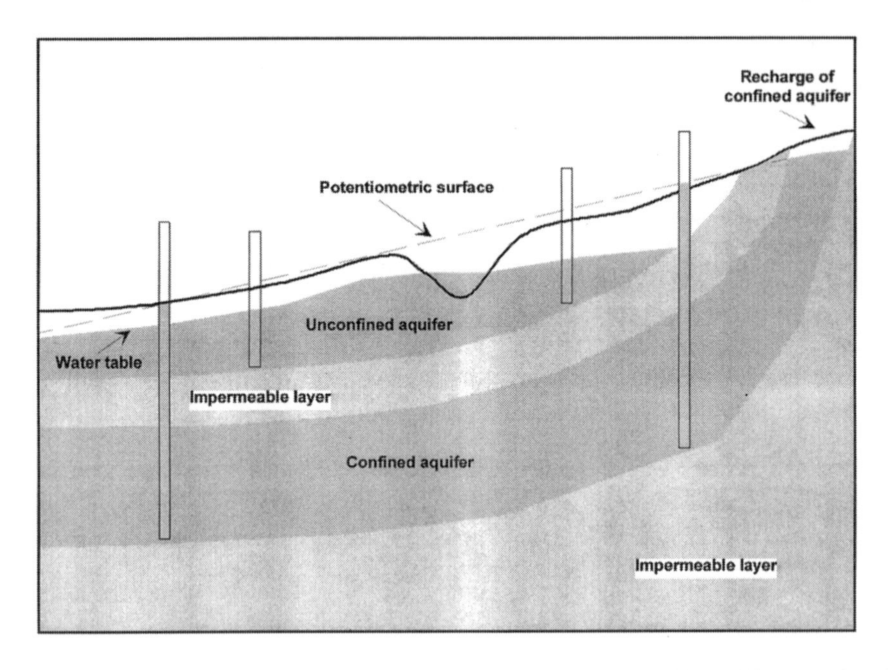

FIGURE 2.2 Unconfined aquifer and its water table; confined aquifer and its potentiometric surface.

Water found in deep aquifers may have been stored over millions of years and is sometimes referred to as "fossil water." The natural rates of recharge to these deep aquifers, when recharge occurs at all, are quite low (Lloyd and Farag, 1978). Fetter (1994) notes that for practical purposes such aquifers are not recharged and any extractions are irreversible. The extraction and use of water from such aquifers is analogous to the mining of resources such as minerals that do not recur periodically on anything less than geologic time scales. Aquifers in arid regions are frequently characterized by very small rates of recharge that range from a few hundredths of a millimeter per year to perhaps 200 mm/yr (Heath, 1983). Aquifers characterized by either the total absence of recharge or by very low rates of recharge cannot be relied upon as a sustainable source of water supply. The Ogallala aquifer underlying parts of Texas, Oklahoma, and New Mexico is a good example of such an aquifer. The relatively high rates of extraction and use of water from the Ogallala aquifer for agricultural purposes over the past four decades has resulted in progressive increases in pumping depths. In many places the depth to ground water is so great that it is no longer economical to pump. In these areas irrigated agriculture that historically relied on waters from the Ogallala must be converted to dry land farming or other land uses.

Ground water depletions occur when water is discharged from aquifers naturally via seeps and springs, from direct uptake by plants where the water table is in the root zone, and from extractions through wells. The manner in which a ground water basin responds to pumping depends upon whether the aquifer is confined or unconfined. For a confined aquifer, a cone of depression, which originates at the point that water is actually extracted by pumps, will move rapidly through the aquifer. Thus remote parts of the aquifer will be affected and some of the natural discharge will be captured. For unconfined systems, the cone spreads too slowly to affect distant points of natural discharge so that most of the water removed comes from storage. The ease of pumping is related to the capacity of the aquifer to conduct water, its hydraulic conductivity. Aquifers with low conductivities will pass water only very slowly so that wells must be deep to produce adequate supply. The increased depth requires increases in pumping lifts, which translate directly to increased pumping (extraction) costs.

In aquifers that are undisturbed by human activity, recharge tends to be balanced by natural ground water discharge or extractions. This means that water tables in unconfined aquifers and the potentiometric surfaces in confined aquifers remain stable. When this steady state is disturbed by ground water pumping or diversion of customary sources of recharge, water tables and potentiometric surfaces respond accordingly. Thus, for example, in unconfined aquifers the water table rises when the rate of recharge exceeds the rate of extraction and discharge. Conversely, if extractions exceed recharge, water tables will fall, as will surface discharges such as base flow in streams (Figure 2.3). As a general rule, however, rising or falling water tables cannot be sustained indefinitely, and the aquifer will always tend toward a steady-state condition where the rates of extraction and discharge are equal to the rate of recharge. For this reason, the sustainable or safe yield of any aquifer is equal to the long-run average rate of recharge.

Conjunctive Use of Surface and Ground Water

Conjunctive use of surface and ground water may be defined as any integrated plan that capitalizes on the combination of surface and ground water resources to achieve a greater beneficial use than if the interaction were ignored (Morel-Seytoux, 1985). Interactions of this kind occur naturally in alluvial valleys and flood plains, but under present circumstances prudent watershed management often necessitates engineered approaches to enhance the natural processes. Such management of overall water resources often takes the form of storing surface water underground in times of surplus by recharging natural ground water aquifers, thus saving the enormous cost of above-ground storage reservoirs and aqueducts.

Moreover, long-term storage in and passage through a ground water aquifer generally improve water quality by filtering out pathogenic microbes and many, although by no means all, other contaminants (NRC, 1994b). Ground water

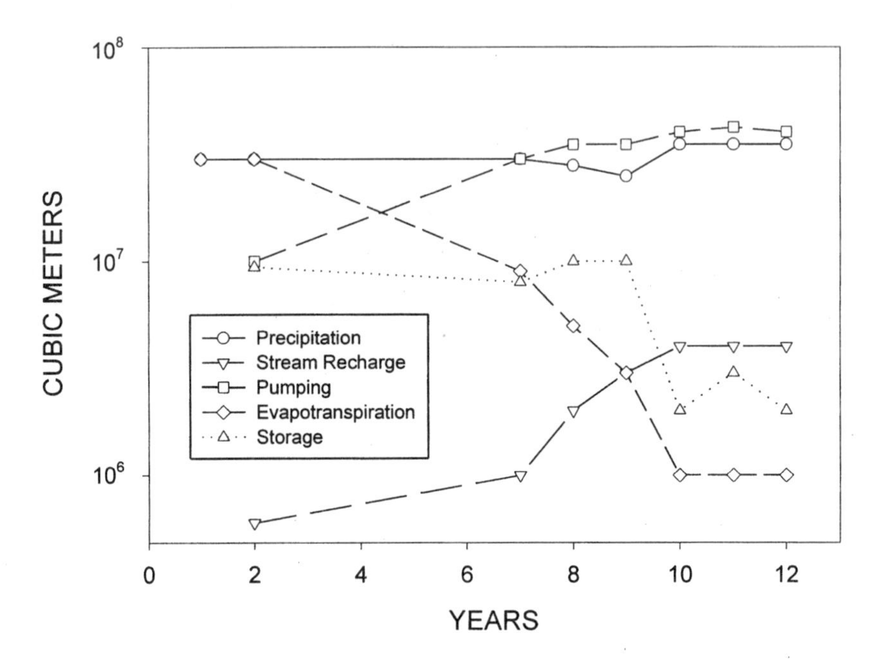

FIGURE 2.3 The effect of pumping on service flows provided by a hypothetical aquifer. With regard to this inventory, the following points are worth noting: (1) Prior to pumping the aquifer, natural recharge equaled natural discharge, and the ground water basin was in a steady state. (2) With the addition of pumping and in the course of withdrawals from storage, net recharge to the aquifer from stream flow increased and reached some maximum value (i.e., part of the stream flow was captured by pumping the aquifer), whereas discharge by evapotranspiration decreased and approached some minimum value (i.e., the amount of plant-viable water was reduced). (3) During the course of the withdrawals, the basin was in a transient state where water was continually being withdrawn from storage. Although not shown, this results in a continual decline in water levels. A new steady state could be achieved by reducing pumping to about $3.8´10^7$ m^3 yr^{-1}, but the steady state would include the reduced stream flow and evapotranspiration. Data for this figure were taken from Domenico and Schwartz, 1990. (Reprinted with permission from John Wiley & Sons, Inc., 1990. Copyright 1990 by John Wiley & Sons, Inc.)

supplies generally are far superior to surface water sources (American Water Works Association, 1990). Indeed, where available, ground water basins afford benefits of storage, conveyance, and treatment that often render the ground water resource preferable to surface water alternatives from the standpoint of health protection, technical simplicity, economy, and public acceptance.

THE ECONOMICS OF GROUND WATER USE

There is a significant and varied literature on the economics of ground water use (see, for example, Burt, 1970; Cummings, 1970; Burness and Martin, 1988; and Provencher and Burt, 1993). Several common principles emerge from this literature, perhaps the most important of which is that ground water is used most efficiently when it is extracted at rates that maximize net benefits (total benefits net of total costs) over time. Costs include the cost of extracting and delivering the ground water and the opportunity, or user, cost. The benefits are determined by the uses to which the water is put.

The costs of extraction are primarily a function of pumping technology (or pump efficiency), the depth from which the ground water must be pumped, and the costs of energy. These costs increase with pumping depth and the cost of energy and decrease as pump efficiency is improved. The cost of extraction also includes the value of the opportunity foregone by extracting and using the water immediately rather than at some time in the future. The user cost is a measure of the economic consequences of pumping now and thereby lowering the water table and increasing costs of extraction for all future periods. The extraction rate in the current period will be efficient only if the potentially higher costs of pumping in the future periods are appropriately estimated. Much economic literature on ground water resources emphasizes that when ground water is pumped in an individually competitive fashion, pumpers have strong incentives to ignore the user cost. In these circumstances pumpers tend to treat ground water as an open access resource, with the result that rates of extraction exceed the economically efficient rate.

The tendency to consider ground water an open access resource when it is exploited competitively underscores the importance of well-defined, clearly enforceable rights to extract ground water. These rights may be assigned to individuals or the citizens of a political entity. They may also be permanent or time limited and subject to change. In instances where rights are not effectively defined and enforceable, the availability of ground water is determined by and subject to the law of capture: whoever taps the ground water first gets to use it. Pumpers have an incentive to extract as much water as possible, subject to the constraints imposed by pumping costs. Incentives to conserve voluntarily are absent, since water not pumped is available to competing users and will not necessarily be conserved for future periods. Thus, competitive pumpers often ignore user costs both because they believe that self-discipline will not effectively conserve supplies for the future and because they believe that the impact of their own pumping on the water table will be small. When the user cost is ignored, the costs of ground water extractions are undervalued and water is extracted too quickly. This contrasts with situations in which ground water is extracted by a single pumper. The single pumper accounts for the user cost simply because he or she will have to bear all of the additional costs of pumping from a lowered water table in the future. In competitive situations regulatory measures can be used to ensure that pumpers account for the user cost. Two common measures are the imposition of a pump tax equivalent to the user cost or the imposition of pumping quotas to ensure that the aquifer is not exploited too quickly (Nether, 1990). Such measures can be imposed by ground water management agencies, and where taxes are employed, the revenues could be used to defray the costs of managing the ground water basin, should that be the most efficient use of the funds.

Defining and enforcing ground water extraction rights and ensuring that rates of extraction are efficient are equally important in decisions to invest in the protection of ground water quality, as well as in programs to remediate or enhance ground water quality. If ground water is subject to the law of capture, then the benefits of protection, remediation, and enhancement investments will similarly be subject to the law of capture. This results in less than optimal investment in the preservation and enhancement of ground water quality, since those investing in such measures cannot be sure they will capture all of the benefits. This fact underscores the necessity of establishing clear and enforceable systems of extraction rights and appropriate regulatory measures before investing in the protection and enhancement of ground water quality.

In the long run, rates of ground water extraction cannot exceed rates of recharge. That is, over time, rates of extraction and recharge will be brought into steady-state equilibrium. When overdrafting occurs persistently, water tables are lowered and pumping costs increase. Finally a point is reached where the costs of extracting ground water exceed the benefits that can be obtained from its use; then pumpers stop extracting and the decline in the ground water table is arrested. Because ground water is extracted and used only when it is profitable to do so, overdraft will be self-terminating and rates of extraction will ultimately be exactly equivalent to the rates of recharge.

It is important to recognize, nevertheless, that ground water overdraft may be economically efficient in some instances. When the benefits of use are quite high in relation to the costs of extraction (including the user cost), overdraft may be efficient for some period of time. In periods of drought, for example, when surface water supplies may be absent or scarcer than normal, overdraft may be efficient. However, even in situations where overdraft is efficient, it will ultimately be self-terminating. Moreover, in assessing the economic desirability of overdraft, we must account for certain adverse impacts, such as land subsidence, salt water intrusion, and deleterious effects on surface water and aquatic habitats

The geological substrate of aquifers differs from location to location, with materials ranging from coarse sediments to fractured rock. Substrates that consist of fine grained sediments such as clays tend to compact when water is removed, resulting in elimination of the pore spaces that previously contained water. Thus removing water reduces the aquifer's water-holding capacity. In addition the land surface may sink when compaction occurs in such aquifers.

This may cause severe disruption of utilities such as sewer and water lines and damage to structures and roads. Subsidence can also cause flooding, particularly in coastal areas. Between 1906 and 1987, land in the Houston/Baytown region of Texas subsided by between 1 and 10 feet, resulting in pronounced flooding of valuable land adjacent to Galveston Bay. When policy-makers recognized the value of remaining ground water in preventing subsidence and concomitant flooding, they formulated a plan to conserve ground water *in situ* by developing sufficient surface supplies to accommodate 80 percent of the projected demand for Houston by the year 2010 (Schoek, 1995). The most dramatic example of subsidence is found in the San Joaquin Valley of California, where land surfaces have fallen up to 40 feet in some areas.

A unique problem associated with subsidence caused by prolonged overdrafting has been the development of sinkholes in some areas of Florida where natural flow patterns in limestone aquifers have been perturbed. Land subsidence generally occurs when aquifer pressure levels are significantly lowered in basins where the substrate is primarily fine-grained material such as clays and silts, which are more compressible than more rigid coarse grains such as sand or limestone and sandstone formations. Subsidence caused by the consolidation of fine-grained material cannot be reversed by artificially injecting additional water into the formation. Subsidence is reversible only in aquifers usually dominated by sands, gravels, or sandstone, which can accept the additional fluids.

Saline ground water is found in aquifers throughout the United States. Ground water depletion may cause intrusion of poorer-quality water into high-quality water supplies. In some coastal regions, particularly in California and Florida, there are serious sea water intrusion problems caused by the attenuation of fresh ground water flows toward the ocean. The *in situ* value of ground water in these cases derives from providing a barrier to salt water intrusion. Overdrafting can depressurize confined aquifers, leading to the intrusion of salt water into portions of the aquifer that formerly contained high-quality water (see Figure 2.4). Salt water intrusion problems are not limited to coastal areas. Problems with saline ground water have been documented in 41 states (Atkinson et al., 1986). A number of methods are available to combat salt water intrusions, including artificial recharge, reductions in extractions, establishment of a pumping trough along the coast, formation of pressure ridges through artificial water injection, and installation of subsurface barriers.

Discharges from unconfined aquifers are the source of about 30 percent of the nation's stream flow (Frederick, 1995). This source of surface water is especially important in sustaining stream flow during dry periods, the so-called base flow. Ground water levels have a direct impact on lake levels and on the amount of freshwater flowing through estuaries to the oceans. Reductions in surface water flows can have adverse impacts on the aesthetic values, recreational potential, and use of surface waterways for transportation.

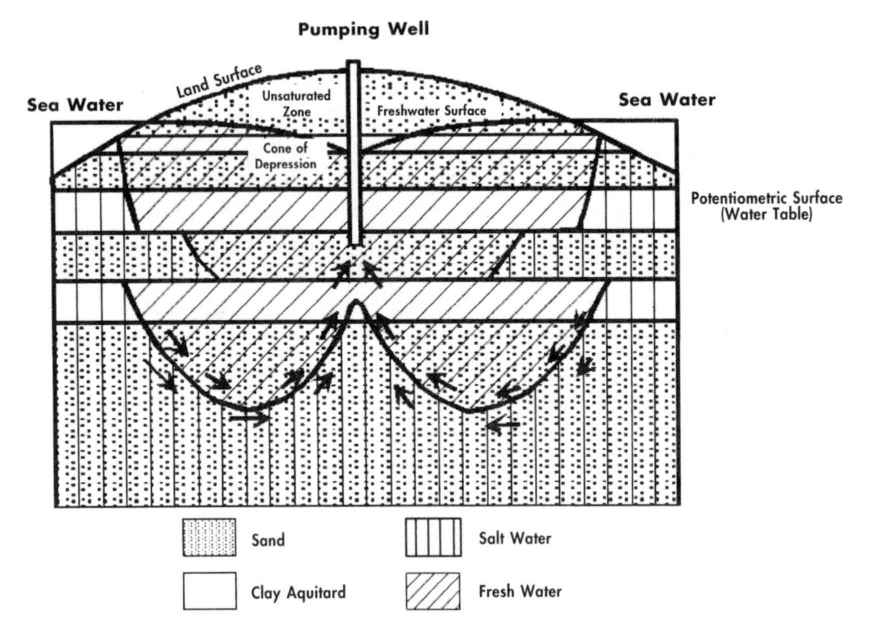

FIGURE 2.4 Salt water intrusion into a confined aquifer.

Surface water flows originating from ground water also support riparian vegetation and play a major role in maintaining wetlands (NRC, 1995). Such support constitutes a vital ecological service. Ground water depletions are known to have eliminated surface water flows altogether in some areas. Many of the flowing streams in Arizona have disappeared because of the overpumping of ground water. High water tables may also support riparian species in areas where surface flows are ephemeral. The ecological services of ground water are particularly dramatic in cases where ground water supports habitat for endangered species. (An example of how ground water drawdown can affect stream flow appears in Figure 2.3.)

The availability of ground water is thus determined by the interaction of geological, hydrologic, and economic factors. The quantities of water available now and in the future depend upon the interaction of recharge and extraction. The cost of obtaining ground water is determined by pumping depths, energy costs, and the cost assigned to the opportunity foregone as a consequence of extracting ground water now rather than later. The value of ground water depends upon both the cost of obtaining it and the willingness of users to pay, and willingness to pay depends crucially on the quality of the water.

GROUND WATER QUALITY

Contamination

Because ground water exists in an environment that includes a mineral matrix and perhaps some organic matter (even living organisms), the quality of the water is controlled by the physical, chemical, and biological processes that interact in the aquifer. Ground water exists in a variety of geological settings, ranging from tiny cracks in otherwise solid rock to the (relatively) large voids between grains of coarse sand or gravel. Geological formations that constitute aquifers differ widely in the rocks and minerals they contain. Some contaminants occur naturally, whereas others are derived from human activities: landfills, agriculture wastes, industrial spillage, and many others (see Table 2.1).

In many areas the greatest threat to the potability of ground water is from contamination by microorganisms such as bacteria and by disease-causing virus particles. The presence of potentially pathogenic microbes (expanding the definition of microbes to include viruses) represents the most serious drinking water contamination problem. The organisms of concern in potential ground water contamination are those that are shed in fecal material, including bacteria, viruses, and protozoan parasites. These organisms are spread via the fecal-oral route. Ingestion of organisms can occur through consumption of contaminated food or water or by direct contact. Organic contaminants are wide ranging and include chlorinated hydrocarbons (e.g., trichloroethylene, carbon tetrachloride), fuel hydrocarbons (e.g., benzene, toluene, xylene), oxygenated compounds (e.g., phthalates and phenols), polynuclear aromatic hydrocarbons (PAHs, e.g., arochlor). Many times the contaminants are mixtures, e.g., gasoline, diesel fuel, and creosote. In some cases, contaminant plumes may cover many square miles of aquifer material (NRC, 1994b).

One of the major differences between surface water and ground water is the time frame for contamination. Contamination in ground water develops slowly, based on migration and flow rates. In addition, once contaminated, ground water takes far more time to assimilate and recover than does surface water. Surface water is generally contaminated rather quickly and has the ability to purge the contaminant in a short period of time. Both natural and artificial cleanup of ground water are lengthy processes because of slower flow rates, slower dilution, and reduced capacity for reoxygenation.

Remediation: An Economic Outlook

Policy-makers cannot select an appropriate treatment technology until they define the final use or disposal location of the water. As they evaluate possible treatment, then, they must consider the water-quality objectives for the receiving waters. They should identify feasible disposal options. Although most areas

TABLE 2.1 Sources of Ground Water Contamination

Category I	Category II	Category III
Sources designed to discharge substances	**Sources designed to store, treat, and/or dispose of substances; discharged through unplanned release**	**Sources designed to retain substances during transport or transmission**
Subsurface percolation (e.g., septic tanks and cesspools) Injection wells Land application	Landfills Open dumps Surface impoundments Waste tailings Waste piles Materials stockpiles Aboveground storage tanks Underground storage tanks Radioactive disposal sites	Pipelines Material transport and transfer
Category IV	**Category V**	**Category VI**
Sources discharging as consequence of other planned activities	**Sources providing conduit or inducing discharge through altered flow patterns**	**Naturally occurring sources whose discharge is created and/or exacerbated by human activity**
Irrigation practice Pesticide application Fertilizer applications Animal feeding operations De-icing salt applications Urban runoff Percolation of atmospheric pollutants Mining and mine drainage	Production wells Other wells (nonwaste) Construction excavation	Ground water-surface water interactions Natural leaching Salt water intrusion, brackish water

SOURCE: Office of Technology Assessment, 1984.

favor beneficial use of treated ground water, in certain instances disposal would be cost-effective. Disposal options include placement in evaporation ponds (probably limited to the southwestern United States), deep-well injection, and ocean discharge (limited to coastal areas).

As indicated, appropriate treatment depends on both the types of contaminants and the intended beneficial uses of the renovated ground water. Treatment technologies commonly in use today and their effectiveness for removing specific contaminants and their associated costs appear in Table 2.2. The costs

TABLE 2.2 Effectiveness and Typical Costs of Treatment for Water Containing Various Classes of Contaminants

Treatment Process	Inorganic Compounds		Organic Compounds			
	TDS	NO_3^-, SO_4^{2-}	Volatile (TCE, PCE)	Nonvolatile (DBCP)	Cost Range ($/acre-foot)	Facility Capacity Range (MGD)
GAC	–	–	–	+	150-110	2-12
Air stripping	–	–	+	–	130-50	0.5-7
Ion exchange	–	+	–	–	130-60	1-15
Reverse osmosis	+	+	–	–	400-250	1-6
In situ bioremediation	–	–	+	–	Varies	Varies

SOURCE: Orange County Water District, 1996.

shown are only for the indicated treatment process, that is, the costs do not include costs of extraction wells and collection systems or conveyance and disposal costs. Note that the cost for each treatment process depends upon the size of the facility. Thus the unit treatment cost will decrease as the size of the facility increases. Bioremediation is not included in Table 2.2 because the costs are controlled by the site at which the technology is being employed.

Aquifer Remediation

A recently published National Research Council study dealt with the cleanup of contaminated aquifers and ground water (NRC, 1994b). That study was motivated by the need to assess critically the feasibility of restoring ground water quality at hazardous waste contamination sites, considering the limitations of present technology as well as foreseeable advances in methodology. The need for such an assessment stemmed in turn from disappointment in the slow rate of progress in hazardous waste site remediation and its burgeoning cost.

The NRC committee found that the general frustration with the slow progress and rising costs in hazardous waste site (sometimes referred to as Superfund site) remediation is indeed justified. Only a very few sites have, in fact, been renovated successfully, while efforts at many others have been hampered by inept planning, unrealistic objectives, ponderous decision-making processes, and conflicts among the various stakeholders. Nonetheless, at the bottom of the problem are intrinsic technical difficulties that would be hard to counter even with near-perfect planning procedures in an ideal institutional setting. Where complete restoration remains elusive, it may be prudent simply to contain the contamination after removing the portion of the contaminant mass that is amenable to cleanup.

In the face of these newly perceived difficulties, the task of restoring ground water quality seems considerably more daunting than when the Resource Conservation and Recovery Act (RCRA) and Comprehensive Environmental Response, Compensation, and Liability Act (CERCLA) programs were instituted. Estimates of total costs of cleanup in the range of hundreds of billions of dollars raise the question of whether all contaminated ground water can and should be remediated to the strictest criteria: that is, pristine conditions or health-based standards. This in turn raises the question of the long-term economic and resource impacts of permitting ground water resources to deteriorate in quality. Furthermore, it is necessary to take into consideration the observed tendency of subsurface contamination to become more intractable the longer it is left in place, so that long-term contamination may be virtually irreversible.

Hydrologic Uncertainty

Hydrologic uncertainty results from the heterogeneity of natural systems and from data inadequate to characterize and model the systems accurately. Uncertainty arises regarding both the quantity and quality of ground water systems. Uncertainties related to ground water flow include insufficient or erroneous data from imprecise measurements and observations, sampling errors, or statistical errors; inappropriate model assumptions; and inadequate characterization of subsurface hydrology. Uncertainty regarding quality arises from lack of information on both the fate of the contaminants in the subsurface, and their health effects.

Additional uncertainties concern the role of ground water in providing ecological services. Ground water supports microbial habitats in the subsurface and surface flows that sustain riparian habitats. Connections between ground and surface waters are better defined in theory than in application.

Mathematical models of ground water systems have been under development for decades, but data are rarely if ever adequate to allow accurate prediction of subsurface dynamics in three dimensions. Model uncertainty stems from shortcomings in current theory or failure of models to incorporate the elements of current theory, scarcity of field data for model calibration, inadequacies of computer capacity for modeling complex systems, and failure to incorporate operational constraints into models (Anderson and Burt, 1985).

RECOMMENDATIONS

This review of hydrological concepts, ground water quality, the influence of societal activities on ground water quantity and quality, and ground water treatment scenarios suggests the following conclusions regarding implications for ground water valuation.

- **Decision-makers should proceed very cautiously with any actions that might lead to an irreversible situation regarding ground water use and management. Ground water depletion, for instance, may often be irreversible. Some aquifers (e.g., the southern edge of the Ogallala) do not recharge in useful time scales, and thus any extractions constitute a form of mining. In other cases the length of time needed for natural recharge of deep aquifers where ground water removal rates are high leads to a continual reduction in stock that will not be replenished in short time frames. Moreover, overdrafting can sometimes lead to a collapse of the formation permanently reducing the aquifer's storage capacity.**
- **Decision-makers should also be cautious regarding contamination of ground water. Restoration of contaminated aquifers, even when feasible, is resource intensive and time consuming. Restoration methods are uncertain and unlikely to improve significantly in the near future. As a result, it is almost always less expensive to prevent ground water contamination than to clean up the water.**
- **Ground water often makes significant contributions to valuable ecological services. For example, in the Southwest, many flowing streams have**

been eliminated by overpumping. Because the ground water processes that affect ecosystems and base stream flow are not well understood, combined hydrologic/ecologic research should be pursued to clarify these connections and better define the extent to which changes in ground water quality or quantity contribute to changes in ecologic values.

- **Ground water management entities should consider appropriate policies such as pump taxes or quotas to ensure that cost of using the water now rather than later is accurately accounted for by competing pumpers.**
- **Because ground water resources are finite, decision-makers should take a long-term view in all decisions regarding valuation and use of the resources.**

REFERENCES

American Water Works Association. 1990. Water Quality and Treatment. Blacklick, Ohio: McGraw-Hill.

Anderson, M. G., and T. P. Burt. 1985. Hydrologic Forecasting. New York: John Wiley and Sons.

Atkinson, S. F., G. D. Miller, D. S. Curry, and S. D. Lee. 1986. Salt Water Intrusion: Status and Potential in the Contiguous United States. Chelsea, Mich.: Lewis Publishers.

Burness, H. S., and W. E. Martin. 1988. Management of a tributary aquifer. Water Resources Research 5(24):1339-1344.

Burt, O. R. 1970. Groundwater storage control under institutional restrictions. Water Resources Research 6(6):1540-1548.

Cummings, R. G. 1970. Some extensions of the economic theory of exhaustible resources. Western Journal of Economics 7(3):201-210.

Domenico, P. A., and F. W. Schwartz. 1990. Physical and Chemical Hydrogeology. New York: John Wiley and Sons.

Fetter, C. W. 1994. Applied Hydrogeology. New York: Macmillan College Publishing.

Frederick, K. D. 1995. America's water supply: Status and prospects for the future. Consequences 1(1):14-23.

Heath, R. C. 1983. Basic ground-water hydrology. Water-Supply Paper 2220. U. S. Geological Survey.

Lloyd, J. W., and M. H. Farag. 1978. Fossil ground water gradients in and regional sedimentary basins. Ground Water 16:388-398.

Morel-Seytoux, H. J. 1985. Conjunctive use of surface and ground water. Pp. 35-67 In Artificial Recharge of Ground Water. T. Asano, ed. Chapter 3. Boston: Butterworths Publishers.

National Research Council. 1994a. Ground Water Recharge Using Waters of Impaired Quality. Washington D.C.: National Academy Press.

National Research Council. 1994b. Alternatives for Ground Water Cleanup. Washington, D.C.: National Academy Press.

National Research Council. 1995. Wetlands: Characteristics and Boundaries. Washington, D.C.: National Academy Press.

Nether, P. A. 1990. Natural Resource Economics: Conservation and Exploitation. New York: Cambridge University Press.

Office of Technology Assessment. 1984. Protecting the Nation's Groundwater from Contamination, OTA-O-233. Washington, D.C.: U.S. Congress.

Provencher, B., and O. R. Burt. 1993. The externalities associated with common property exploitation of groundwater. Journal of Environmental Economics and Management 24(2):139-158.

Schoek, J. M., ed. 1995. City cuts use of depleted ground water. The Ground Water Newsletter 24(12):6.

3

A Framework for the Valuation of Ground Water

This chapter provides a conceptual framework for valuing ground water resources that in turn provides a basis for evaluating the tradeoffs that occur whenever there are competing uses for the ground water resources. For example, continued use of ground water as an input into agricultural production implies that less ground water is available for municipal purposes. The "correct" or economically efficient allocation of a scarce resource such as ground water among competing uses depends in part on how the service flows are valued from each use of the resource.

The framework proposed in this chapter is based on an overall economic valuation approach that integrates the hydrological and physical components of the valuation problem. More specifically, the framework links changes in the physical characteristics (quantity and quality) of ground water resources to changes in the level of services or uses of the resources and finally, to how society values the changes in the services or uses. It is in establishing the connections among the changes in quantity/quality of ground water and the changes in service flows that collaboration among researchers (e.g., engineers, hydrogeologists, and economists) is most essential. How society values the changes in service flows or uses is primarily an economic valuation problem; its outcome is influenced by institutions and individual tastes and preferences.

The steps involved in developing this integrative framework, delineated in the following sections, highlight the needed input from both economics as well as other relevant disciplines. Variations of this interdisciplinary framework have been utilized for the valuation of other resources or resource-related assets, including agricultural land resources, air quality, recreational resources, and wetlands. Boyle and Bergstrom (1994) proposed a similar framework for measuring the economic benefits of ground water in a report prepared for the U.S. Environmental Protection Agency. While the impetus for Boyle and Bergstrom's report was the need to incorporate the value of ground water resources when conducting regulatory impact analyses, the framework they developed and the variation of that framework outlined here are applicable to other policies and programs that affect ground water resources.

SOME PRELIMINARIES

Ground Water as a Natural Asset

Ground water can be considered a natural asset. The value of such an asset resides in its ability to create flows of services over time. As discussed in Chapter 1, there are two broad categories of resource services provided by ground water: extractive and *in situ*. The relationships among these are illustrated in the schematic diagram shown in Figure 3.1.

In each time period depicted in Figure 3.1, a ground water stock provides each of the services and is subjected to various influences that affect its quality. Of course extraction and/or addition of water today affects the quantity and quality of stocks tomorrow, and it is critical to incorporate this intertemporal element into the analysis of the valuation problem. Intertemporal issues are related to each of these service flows, and understanding them is fundamental to understanding the overall valuation problem.

The Concept of Total Economic Value

The total economic value (TEV) of ground water is a summation of its values across all of its uses. Sources of values have been classified into use values (sometimes called direct use values) and nonuse values (also known as passive use values, existence values). The use values arise from the direct use of a good or asset by consuming it or its services. For ground water, these would include consumption of drinking water and other municipal or commercial uses. Nonuse values arise irrespective of such direct use. Thus in the economist's jargon the total economic value of a given resource asset includes the summation of its use and nonuse values across all service flows. The notion of total economic value is fundamental to ground water valuation and should enter into management decisions regarding use of water resources. Valuation is a useful tool if the values can help inform decision-makers. The relevant issue is how the TEV of ground water will change when a policy or management decision is implemented.

Prices are often used as a proxy for values. In settings where goods or services (for example, eggs and haircuts) are traded through markets, prices are a good proxy for the value of the last, or the marginal, unit that is traded. As more

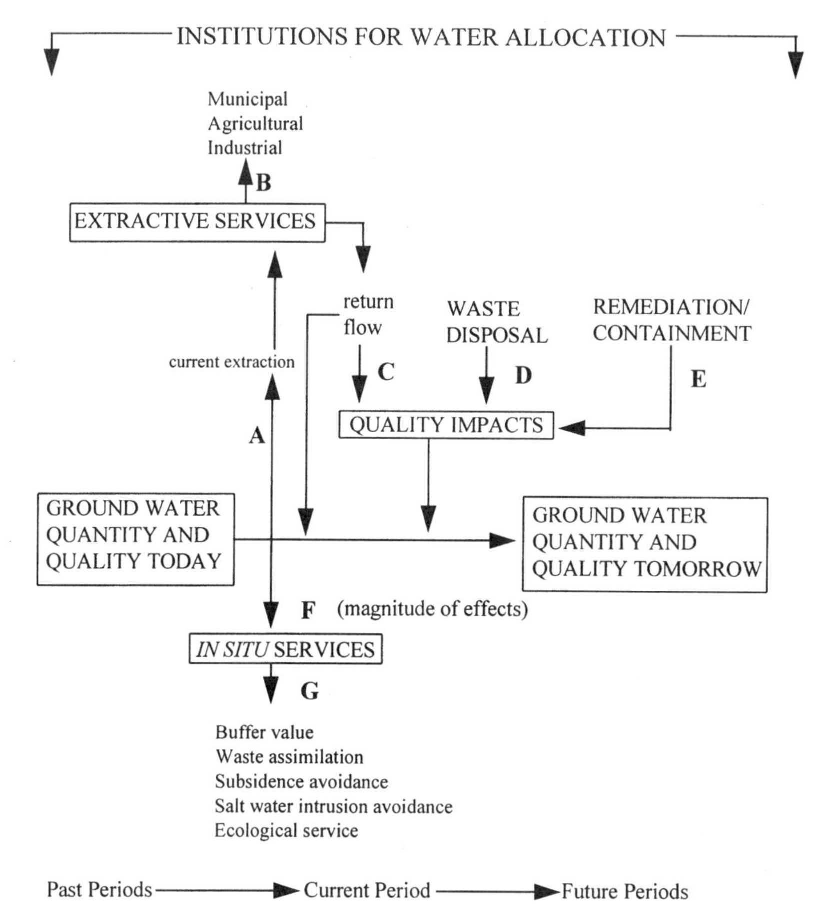

FIGURE 3.1 Ground water services.

and more units are traded, the marginal value continues to decrease, as represented by a negatively sloped demand curve for the good in question. But the total value, represented by the area under the demand curve out to the quantity demanded increases.

Several important points about value have particular relevance to ground water valuation. Values are specified at an individual level, and defining a social value for ground water requires aggregating individual values. There are many possible ways to weight individuals in forming such aggregates, including using unweighted dollar-for-dollar sums. Assigning equal weights across all individuals (i.e., a dollar-for-dollar summation) is a common procedure in benefit-cost analyses of public policies. Use of such a procedure assumes that the current or existing distribution of incomes is socially acceptable. All values derive ultimately from services to consumers, whether these services are consumed directly or through produced goods. In this way ground water, ecosystems, and other environmental resources generate value either directly or indirectly.

Economic valuation methods have concentrated on techniques for assessing particular pieces of the total value puzzle. The easier pieces to value are those associated with identifiable uses such as agriculture, municipal water supply, and other commercial or industrial uses. The examples in Chapter 6 illustrate the noncomprehensive approach to valuing a ground water resource, where the focus has been primarily on valuing ground water resources in their direct use purposes.

Finally, there are no restrictions on why someone values a good. Economic values are anthropocentric notions and are based on situations of choice. The mechanism of choice might be a market or a negotiated explicit or implicit contract or a public referendum. Because this valuation is based on human choices, it does rule out some of what concerns some ecologists and environmentalists who believe that nature inherently has "rights." Therefore the concept of economic valuation does have some limitations in discourse about natural resource policy where the "rights" of nonhuman entities are given significant weight compared to human use values.

Nonuse values are more controversial than use values when it comes to measuring and validating them. Some of the techniques presented in Chapter 4 suggest ways to quantify the nonuse values as part of measuring the total economic value. The issue of how to model and measure nonuse values cannot be totally separated from the measurement of use values. And as Freeman (1993b: 161-162) indicates:

> economic theory gives unambiguous guidance only on defining total values as compensating income changes for changes in a resource. The question of whether non-use values, however defined, are positive takes on meaning only after some decision has been made about what use values measure, since non-use values are simply total value minus whatever has been called use value... Ultimately we want to be able to measure total value. Any distinction between use and non-use values is itself useful only if it helps in the task of measuring total value.

Although there is no *a priori* agreement on when nonuse values are likely to be significant, economists often suggest that one factor would be whether the resource in question is sufficiently unique, has no close substitutes, and has a low price elasticity of demand. In certain locations ground water could satisfy these requirements, particularly if it is valued as a source of "pristine" water. Even in situations where the ground water by itself does not produce much in the way of nonuse value, it may contribute to habitat for endangered species, which has significant nonuse value. In such a case, the derived value of the ground water will include this value as well.

Institutions and Decisions

The value of ground water, that is, its ability to produce valuable service flows to people, is increased when any given amount of water is allocated efficiently across potential water uses. Water is efficiently allocated when the increment to value that could be obtained from using a little more water in any one way (called the marginal value of water in that use) is the same across all uses of water. To understand this concept, assume that such a balance does not exist. For example, suppose that one use (say, an industrial process) could generate $100 in incremental value if a little more water is used there, while another use (say, agricultural production) would lose only $50 in value if a little water is removed from that use. Then transferring a unit of water from the agriculture to the industrial use increases the total value of water services by $50.

Any inefficiencies in the allocation of ground water across uses or quality will lower the value of the ground water. Poor quality will also reduce its value to users. The value of ground water, then, is intimately tied to institutions that govern how it is allocated (or misallocated) and protected in the current period, through time, and according to its quality. That is why, in Figure 3.1, institutions are depicted as a set of overarching influences that govern the value of ground water.

One possible institution for allocating ground water is a set of private water markets. Economists have shown that if these markets are organized in a particular fashion, then the allocation of water among uses will be efficient. Of course the conditions underlying this result would be difficult to meet. Instead of a market system, there presently exists a complex web of water resource institutions that vary from jurisdiction to jurisdiction. These institutions are described in more detail in Chapter 5. The analysis of how these institutions promote or mitigate inefficient water allocations is an important but difficult task that is beyond the scope of this committee's investigation.

The total value of ground water is increased if it is efficiently allocated. Identifying this efficient allocation depends on measuring the incremental value of water in alternative uses and the incremental value of improvements in water quality. When there are large gaps between these incremental values across uses, then economic well-being is enhanced by altering the allocation. If the incremental costs and benefits of changing water quality are greatly different, then economic well-being is enhanced by improving water quality or perhaps by allowing a lower level of water quality for some uses.

Consider first the allocation decision. In Figure 3.2 the horizontal axis shows the quantity of water that can be allocated to either of two uses: in-stream flows, which provide ecological services; or landscaping, which provides aesthetic values that can also be considered a part of ecological services. At the right edge of the diagram, at point Q, all water is allocated to in-stream flows, and none goes to landscaping. The vertical axis measures incremental values. The line

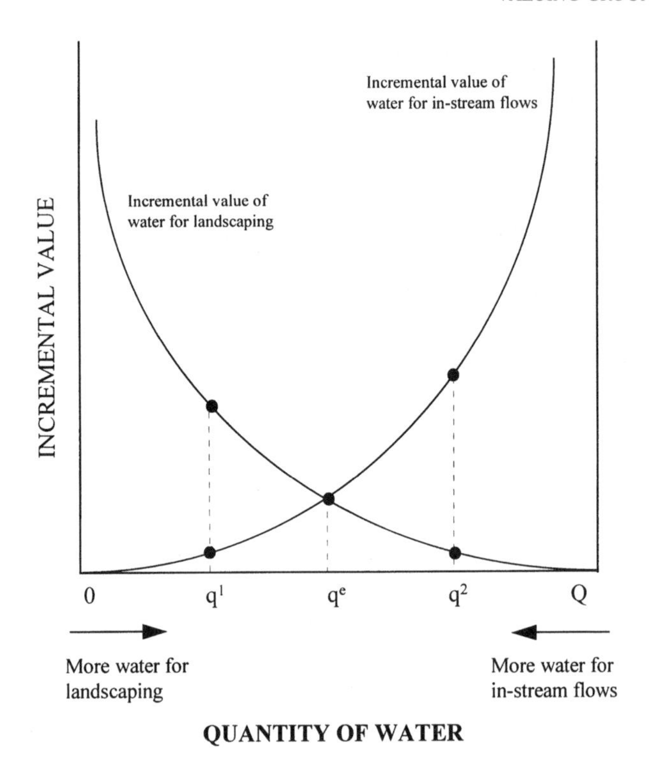

FIGURE 3.2 Allocation decisions.

sloping downward to the right depicts the additional value one could obtain from having more water for landscaping. It is initially very high, since some plants and trees and degree of green is highly valued, but as more and more water is used in this fashion, the additional value that can be obtained falls. Similarly, the incremental value for ecological services is initially very high, since some water in streams sustains basic biological functions, but it, too, falls as more water is allocated to this purpose.

The efficient allocation of ground water balances incremental values across the two uses. This is shown at a point q^e in Figure 3.2. At point q^1, more water should be allocated to landscaping, while at point q^2, stream flows should be increased.

Currently, there is minimal information on the value of ground water in many of its alternative uses. Much has been written regarding municipal, agricultural, and industrial uses and the inefficient institutions that artificially depress the value of ground water in agricultural uses in the American West. The idea is that there are large gaps between incremental values of water across these uses. Although extractive uses have been widely studied, almost nothing is known

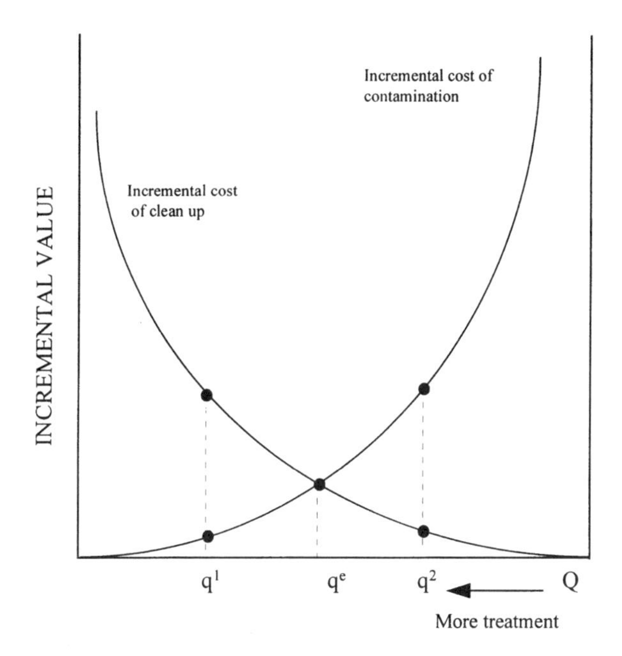

FIGURE 3.3 Quality decisions.

about *in situ* ground water services and their values. Even the incremental value of many municipal water uses, such as landscaping, is not well understood.

Consider now the quality decision. Figure 3.3 on quality decisions shows a diagram similar to the last one but with a different interpretation. Suppose a remediation decision is to be made. The horizontal axis shows the degree of contamination of a ground water stock, with increasing contamination to the right. The contaminated stock is at quality Q. Improving ground water quality via treatment is, as explained in Chapter 2, a costly process. The incremental costs are shown as moving upward to the left. Information is available about the cost of alternative technologies for cleaning up ground water; for example, a recent NRC report, Alternatives for Ground Water Cleanup (NRC, 1994), is directed to this issue. However, this information does not address whether or how much to remediate or treat.

The decision process requires value information. The relevant value is the incremental value of enhanced water quality, shown by the line increasing to the right in Figure 3.3. The efficient quality decision lies at quality level q^e. Treatment levels between Q and q^e such as q^2 (Figure 3.3) represent less than economically optimal treatment since the benefits from additional treatment outweigh the costs. If water is left untreated (point Q), gains from treatment outweigh costs up to a quality level q^e. Treatment to achieve qualities greater than that at q^e (as, for example, to q^1) is excessive since benefits are outweighed by the costs.

These simple examples depict how valuation measures can be used in decision-making. Obviously, these are highly stylized. In practice these nice smooth curves do not exist, and there are lumpy, nonincremental decisions to reach. But the basic point remains. The valuation framework described later in this chapter is of interest not in its own right as an academic exercise but rather as part of a decision-making process. One set of relevant decisions involves ground water management decisions within a given institutional structure. Values can also be used for institutional reform, to reduce systematic inefficiencies.

A major question concerns the availability of value information for use in decision-making. When assets and their services are exchanged in organized markets, there is an observable link between the asset's value and the values of the services that the asset provides (Kopp and Smith, 1993). For example, the current value of a commercial building (an asset) can be determined in the real estate market. Its value can also be appraised by examining the present discounted value of the stream of net incomes realized over time (as a result of annual rentals), plus any residual value. If the building is damaged, the value of the asset is reduced precisely because the present discounted value of the stream of net rental incomes is reduced. In this case the existence of organized markets provides information on how society values the asset.

In the case of ground water, however, other nonmarket institutions govern its use. Neither the asset nor its services are traded on well-organized markets. Thus, no ready source of information automatically provides a connection from service values to asset values that would be similar to the information the market provides.

The valuation process that managers and policy-makers undertake should, in theory, be similar to the valuation process the market provides. One needs to know the time stream of services ground water supplies, and the values that society places on these services. These values are not straightforward for two reasons. First, there are many services for which information on individual values is not readily obtained. For example, because ecological services fall outside of markets, they call for specialized valuation techniques. Second, values are defined and measured at the individual level. The difficulty comes in deciding how these should be aggregated across people.

Because services exist across time, an appropriate discount rate must be used to determine the *present* value of this stream of annual service values. This is rather like the problem of adding up values across people: now we need a way to add up value across people alive at different points in time. The market rate of interest serves as an approach for money assets, but things are not quite so simple for natural assets provided through public institutions. This is a complex question tied to issues of intergenerational equity.

The next step in the process is creating a link between the management decision to be implemented and the resulting changes in the time path of services the ground water stock will provide. Considerations must go beyond mere description of the services already used in one state of the world; they must also involve predictions of future services.

Allocation over Time and Discounting

The provision of benefits or services over time requires that consumers and other users "trade off" benefits (or costs) in one period, such as the present, against benefits in a different time period. In other words, consumers/users of the resource must balance the desire for current consumption against a desire for consumption in the future. All else being equal, people would rather consume a unit of a good today rather than wait to consume it in the future, say in a year's time. Having a unit of income today is worth more than having the same unit of income a year in the future.

As indicated earlier, ground water is considered to be a common limited resource that can be used at different rates over time. Water managers must be concerned with the preferences society has for using that limited resource and the manner in which the ground water will be mined under alternative institutional arrangements. Both of these concerns lead to notions of intertemporal use and discounting.

The traditional criterion used to address the problem of use rates over time is to compare the net benefits (benefits minus the costs) received in one period with the net benefits received in another period. The concept that allows for making this comparison is called present value, which explicitly incorporates the time value of money. The present value of a onetime net benefit received a year from now is computed as

$$(\textit{net benefits in year } 1) / (1 + r),$$

where r is the appropriate interest rate. This process of calculating present value is known as discounting, and r is referred to as the discount rate.

Using the notion of discounting, we can determine an economically efficient allocation of a resource over time: an allocation of a resource across n periods is efficient if it maximizes the present value of net benefits that could be received from all possible ways of allocating the resource over the n periods. For water managers, the challenge is to balance the current and subsequent uses of the ground water stocks by maximizing the present value of the net benefits derived from the limited resource. Knowing the total economic value of the ground water is crucial for determining the net benefits.

Scarcity imposes an opportunity cost, which economists refer to as a marginal user cost. Greater use of the resource today diminishes future opportunities for use, so the marginal user cost is the present value of these foregone opportunities. Using ground water for watering lawns and agricultural purposes may not be appropriate under conditions where drinking water supplies to future generations are denied but may be wholly appropriate in situations with sufficient supplies of water. Failure to take higher scarcity value of water into account will lead to extra costs to society by imposing extra scarcity on the future. Conversely, overconservation in areas with sufficient supplies will impose additional costs on society today.

Allocation of the ground water resource over time is affected by the discount rate. The higher the discount rate, the greater the amount of the resource that will be allocated to the earlier periods. Higher discount rates skew consumption and use toward the present because they give less weight to future net benefits. The methodology for choosing an appropriate discount rate is a matter of continuing debate: "after a lot of time trying to discover an unassailable definition of the social rate of discount, economists are beginning to decide that a totally satisfactory definition does not exist" (Page, 1977). The proper rate depends, in part, on the context of the decision being analyzed (Lind, 1990) but the role of the discount rate is to ensure that scarce resources are allocated efficiently over time.

Issues of whether resources are allocated fairly over time are different, albeit important, issues. In *Sustaining Our Water Resources* (NRC, 1993a), Brown Weiss notes that: ". . . the withdrawal of ground water in excess of recharge rates to supply potable drinking water or rapid withdrawal of water from nonrechargeable aquifers, will cause conflicts between immediate satisfaction of needs and long-term maintenance of the resources." Brown Weiss notes further that in these cases "means need to be developed to reconcile intergenerational concerns with the demands of the living generation."

Issues of whether allocations over time are fair are difficult to address through the selection of a discount rate. Recent literature has raised questions about the applicability of cost-benefit analysis as currently practiced to deal with intergenerational issues. Smith (1988) suggests that many resource problems we face today, including depletion of ground water resources, "stretch the conceptual basis for benefit-cost analysis well beyond the bounds for what it was intended—a single generation borrowing from itself." Page (1988) suggest a fundamental change in how economists evaluate allocation issues that span many generations. In his view the question is not simply one of selecting the appropriate rate of discount, but of basing policy decisions on an intergenerational social choice rule, according to what society considers "fair." In his earlier writings, Page (1977) argues for preserving the opportunities for future generations as a common sense minimal notion of intergenerational justice. Preserving these opportunities is critical in settings where there are irreversibilities and a large degree of uncertainty with respect to both the size of the resource stock and the future demands on the resource. These issues of fairness are distinct from issues of allocative efficiency and, in general, selection of discount rates should be guided by considerations of efficiency while issues of fairness should be resolved in other ways.

Role of Economic Uncertainty

Economic uncertainties occur at both micro and macro scales. The value of a particular ground water supply that supports an extractive use may be influenced by events at local, regional, national, or international levels. For example, the prices of water-intensive commodities such as cotton or copper are affected by price supports and international markets. Changes in commodity prices affect the value of the ground water used to produce those commodities. Farm policies also have an impact on ground water use.

Economic uncertainty is commonly related to lack of data with which to predict human behavior across time and space. Economic uncertainties relative to nonmarket goods and services are even more substantial, because outside of a market there is no documentation of monetary value. Various techniques have been developed to estimate monetary value, but certain values may remain hidden, and there are multiple sources of error in these techniques as well. These techniques and their flaws are discussed in greater detail in Chapter 4.

Externalities and Ground Water

The valuation of ground water involves considerations of external effects inflicted upon ground water, such as the ecosystem side effects incurred when ground water is extracted or contaminated, and the effects that ground water extraction decisions have on ground water availability and cost.

The decisions of any number of consumers and firms may alter ground water quality in unintended ways. Some of these result from point sources of pollution, where a known, identified source is contributing to the problem. We can, at least in principle, measure the quantity of emissions from point sources. Standard approaches for controlling these problems are available, as will be discussed below. Nonpoint source pollution is generated from farms, residences, and urban runoff—a diffuse set of sources such that measurement of emissions from any single source is impractical.

Ground water contributes services to the aquatic ecosystem that individual extractors are not likely to take into account. Contamination of an aquifer may lead to surface water contamination, and depletion may change wetlands, affect water tables, cause land subsidence, and so on. A host of effects greatly complicate the valuation problem; additional examples are in Chapter 2.

Certain externalities arise when one firm's pumping causes other firms' situations to change. These open access resource problems have basically two types of effects. First, pumping may decrease pressure in the aquifer, implying that the total amount of ground water available to all users is reduced. Second, an increase in pumping today increases the pumping costs for all users.

SERVICES PROVIDED BY GROUND WATER

This section offers a brief overview of the different services that ground water resources typically provide (see also Tables 1.4 and 1.5). It also discusses information required to establish the values of such services and how they are affected by changes in ground water policy and/or management.

Extractive Uses

Extraction in excess of net recharge in the current period, as depicted in Figure 3.1 by arrow A, will reduce ground water stocks in the future. Water managers need information to assess how the cost of extraction and distribution is altered by changes in ground water stocks and hydrogeological information to assess how given pumping rates will alter the pressure head in the future. Of course the influence of pumping on future stocks and their quality is a complex issue of hydrogeology and chemistry, since recharge rates, the quality of the recharged water, and aquifer capacity all are involved.

The extractive services consist of municipal, agricultural, and industrial uses of water. Clearly, the efficient allocation of water to alternative uses requires information on relative values in these uses. The municipal uses include direct human consumption, for which strict quality criteria must be met, and a host of other uses with lesser demands on water quality, such as street cleaning, washing cars, and water used for landscaping private residences, parks, and golf courses. Deciding how to value changes in the quantity or price of water for these municipal uses is fairly difficult. Data exist with which to value water for total household use, but how do people value green lawns relative to other uses? Are watered fairways on public golf courses of high or low priority? Further, the supply of water is one issue, the reliability of this supply another. Many ground water development projects are in fact directed to the latter, thus policy-makers need to give attention to valuing changes in the reliability of the water supply along with valuing changes in the quantity of water supplied period by period.

Of greater methodological difficulty is understanding how quality changes alter value, particularly if deteriorated conditions preclude future uses requiring higher quality standards. What demands are placed on water quality by alternative uses? Economists have devoted considerable attention to determining the value of protecting the quality of drinking water from various contaminants. This research is not without controversy, and the committee addresses some of the issues below. But experts also disagree on the health implications of ground water contamination, and the public's perception of the state of this knowledge is even more variable.

Agricultural and industrial uses have a wide variety of water-quality needs attached to them, and the relevant issue is the cost of supplying a sufficient quantity of ground water of suitable quality. The values are fairly straightforward to measure conceptually: the use of ground water contributes to the making of products, and the incremental contribution of water to the value of production measures ground water value in these uses. But of course policy-makers need information from various sources to undergird these measurements. In particular, industrial process engineers or agricultural production specialists might help determine how water quality and quantity changes will affect production. Alternatively, water managers might employ a statistical approach. Arrow B in Figure 3.1 involves interaction between economic and engineering information and, regarding human uses, may involve input from psychometricians (a person skilled in the administration and interpretation of psychological tests), and health experts, as well.

It should also be noted that ground water extraction can be influenced by return flows and their associated quantity and quality. Naturally, this depends on the uses to which ground water is put and on a host of biological, chemical, and hydrological factors. Thus several types of information are needed to elucidate Arrow C in Figure 3.1; such information could be based on input from hydrologists, chemists, soil scientists, and so on.

Further, ground water is subject to pollution from waste disposal and efforts to mitigate such effects. These influences, represented by Arrows D and E in Figure 3.1, are the province of all the current work on ground water contamination, fate and transport of pollutants, movement of pollution within aquifers, effectiveness of alternative remediation or containment efforts, and so on.

Ground water systems are interrelated with surface water systems. Therefore, in the taxonomy defined in this report (see Table 1.3), ecological services are a subcategory of *in situ* services. Understanding of the linkages among ground water resources, wetlands, and lake and stream levels is a complex task for hydrologists, geologists, and aquatic biologists. This information is needed to determine the magnitudes of effects depicted by Arrow F in Figure 3.1. Surface water provides a number of ecological functions, including filtering and processing of pollutants and providing habitat for a wide variety of species, both directly aquatic and terrestrial. The importance of chemists and ecologists is self-evident. Establishing the connections indicated by Arrows F and G in Figure 3.1 is thus a multidisciplinary task.

Ground water contributes notably to many surface water services (see Table 1.5), notably, recreational services. Water in parks makes them more valuable, and swimming, fishing, boating, bird-watching, and a host of other activities either require water or are enhanced by it. There are a variety of methods for measuring recreational values.

In Situ Services

The mere presence of ground water in an aquifer provides a number of services referred to as *in situ* services. First, to some extent, waste products can be added to ground water and their potentially harmful impacts can be mitigated. This assimilative capacity can be thought of in terms of reductions in the cost of other forms of waste disposal or treatment. Obviously, chemists and biologists would determine the capacity of ground water to provide these services (Arrow F in Figure 3.1), and economists and engineers would determine the cost savings implied (Arrow G).

Second, ground water provides structure to the geologic environment. If ground water is extracted, subsidence can occur. The degree to which this happens in any given circumstance is the province of geologists and geotechnical engineers. Civil engineers can assess resulting effects on buildings and infrastructure by direct damage or flooding. The primary economic measure of loss is the dollar value of damage in lost property value or replacement cost for infrastructure. To the extent that the exact degree of subsidence and associated damage is uncertain for given amounts of extraction, economists must assist in analyzing plans for ground water extraction (Tsur and Zemel, 1995).

Very similar to subsidence is the role of ground water stocks in coastal areas in avoiding salt water intrusion. At low levels of stock, reduced hydraulic pressure can allow salt water to invade a coastal aquifer. The extent to which this might happen and at what level of stock is uncertain, but hydrogeologists or engineers can supply some information. The loss in services of ground water is then a matter of the resultant changes in the salinity of ground water. Tsur and Zemel (1995) offer an economic analysis of optimal response to uncertainties in this area.

Ground water also provides a buffer, or insurance service, when managed conjunctively with surface water stocks. Since surface water supplies can fluctuate, ground water acts as important insurance to smooth overall supplies. In times of low surface supply, ground water can be extracted relatively more heavily to augment total supply, and in times of abundant surface supply ground water extractions can fall, allowing the stock to replenish by recharge. Tsur and Graham-Tomasi (1991) have found that this buffer value can be significant. In one example, buffer value constituted 84 percent of the total value of the ground water stock, meaning that if this value were ignored, ground water would be seriously undervalued.

THE CONCEPTUAL FRAMEWORK

This section summarizes the steps involved in an economic analysis of ground water value, noting both the limitations of economic techniques and the inherent uncertainties associated with this task.

Measuring "Values"

Any empirical analysis requires that some preliminary decisions be made regarding the scope of the research. For the ground water valuation problem, this means deciding what value to quantify. The economic concept of value, introduced earlier in this chapter, is the cornerstone of this conceptual framework and is grounded in neoclassical welfare economics. The basic premises of welfare economics are that all economic activity is aimed at increasing the welfare of the individuals in society and that individuals are the best judges of their own welfare. Each individual's welfare depends upon the consumption of private goods and services as well as the consumption of goods and services provided by the government and the consumption of nonmarket goods and services. The latter might include service flows from resources, such as opportunities for outdoor recreation, maintaining wildlife habitat, and visual amenities. Thus it follows that the basis for deriving measures of the economic value of changes in a natural resource, such as ground water, is its effect on human welfare.

The economic theory for measuring changes in human welfare was initially developed for goods and services exchanged in private commodity markets, using observed prices and quantities. Over the past few decades, the theory of measuring economic values has been extended to include nonmarket goods and services. The basis for extending the theory to goods and services that are not traded through private markets is that individuals do substitute among markets as well as use nonmarket goods and services, and this process of substituting reveals something about the values placed on these goods. The value measures are commonly expressed in terms of willingness to pay (WTP) or willingness to accept (WTA) compensation, either of which can be defined in terms of the quantities of a good an individual is willing to substitute for the good or service being valued or in terms of monetary units. (See Freeman, 1993a for a complete discussion of WTP and WTA measures.) The approaches to measuring values discussed in Chapter 4 are attempts to measure either WTP or WTA, when the ground water service flows are not purchased in perfectly functioning markets and have public good characteristics.

A Simple Conceptual Model

Although measuring values involves the use of techniques based on economics, these values must be determined in conjunction with knowledge from other disciplines. For example, estimates of the value of a ground water aquifer in sustaining wildlife habitat must incorporate knowledge of the ecological and hydrological links among the water level of the aquifer, recharge, and the exploited fish and animal species. Estimates of the value of ground water as a source of municipal water depends upon the availability of substitutes, recharge, and other information that water scientists can supply. Lack of knowledge concerning these physical, biological, and/or hydrological relationships is a major limitation to obtaining valid empirical estimates of value of the ground water resource. The conceptual framework presented herein is an attempt to (1) emphasize the importance of economics in valuing ground water resources and (2) make clear that the economic technique and resulting values depend on the underlying relationships that determine the quantity and quality of ground water service flows.

The economic values of the service flows from an aquifer can be viewed as the outcome of three sets of functional relationships; these are functional representations of the flow diagram (Figure 3.1). The first relates some measure of ground water quality/quantity sensitivity to the human interventions that affect it, the second relates the use of the ground water resource and the quality/quantity of the resource, and the third relationship describes government policies and a management plan (Figure 3.4).

The first relationship can be represented as

$$(1) \quad S(t+1) = S[S(t), A(t), z],$$

where $S(t)$ represents a quantifiable measure of the ground water resource (a combination of quality and stock of water), $A(t)$ represents actions taken by people, such as extraction pollution events and remediation, and z represents some random uncontrolled disturbances such as hydrologic events related to net recharge of the aquifer. This relationship shows how the future state of the ground water resource depends on its current state and what is done to it in the meantime. This relationship summarizes a set of purely physical outcomes.

The second functional relationship can be written as

$$(2) \quad A(t) = A[S(t), G(t), Y(S), I(t)],$$

where $A(t)$ is as defined above, representing the set of human activities. $G(t)$ represents a set of governmental policies or management plans. Y represents other background variables, such as the costs of inputs into the production process including labor, capital, and materials that also depend on S, income levels, population, etc. and $I(t)$ is a set of institutional factors that show how decisions are reached and actions taken. This second relationship can be viewed as a decision function that maps the milieu within which ground water decisions are made (government policies, prices of goods and services, income levels, population, the ground water resource, and institutions) into actions (pumping rates, remediation, waste disposal). This shows how changes in any of these factors will alter how ground water is used.

The third set of functional relationships gives the economic value as a function of the uses or service flows. The first relationship in this set shows how the dollar values of the services provided by ground water in any given period depend on those flows and other variables. First, regarding the extractive values, B.

$$(3a) \quad B_{EX}(t) = B_{EX}(A(t), S(t), Y, I(t), z),$$

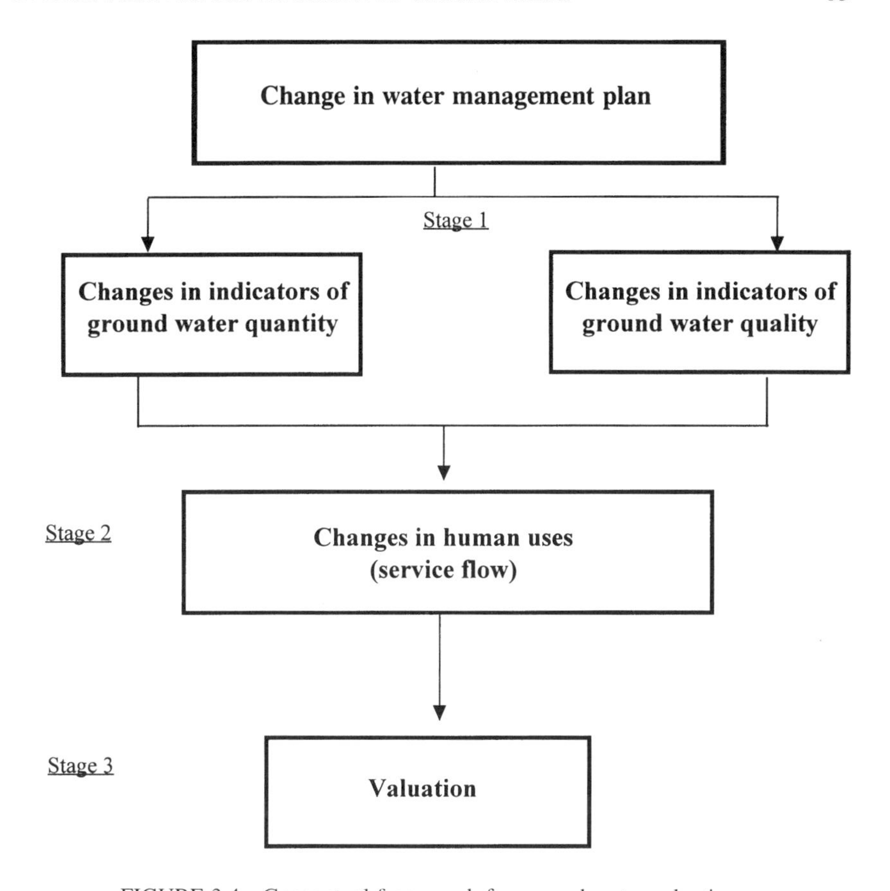

FIGURE 3.4 Conceptual framework for ground water valuation.

which shows that the extractive benefits in the current period depend on the actions taken, the status of the ground water stock, background variables, institutions, and random events, such as rainfall.

Second, regarding *in situ* services, the benefits achieved are given by

(3b) $B_{IS}(t) = B_{IS}(S(t), Y, I(t), z).$

This shows that benefits from *in situ* services are determined by the status of the ground water stock, background variables, institutions, and random events such as salt water intrusion or subsidence events as well as fluctuations in rainfall.

The TEV of the ground water's services in the current period is the sum of the extractive and *in situ* values. Thus, we have

(3c) $\text{TEV}\ (S(t)) = B_{EX}(t) + B_{IS}(t).$

And finally, the value of the ground water stock itself, specified as the present value of the benefits conferred by the service flows that the stock generates can be addressed. Discounting issues and the valuation of the ground water asset will be further discussed in Chapter 4. Here, it can be noted that

$$\text{(3d)} \quad \text{Value of Ground Water} = V(S(t)) = \sum_{t}^{T} \frac{B(t)}{(1+r)^t} .$$

The importance of the discount rate is obvious here.

The set of relationships represented by equations 1 and 2 above are noneconomic in nature and involve a variety of physical, biological, and hydrological processes. The set of relationships represented by equations 3 above depict the integration of the physical and economic sciences. Economists must work closely with other scientists, for an essential input to valuation of *in situ* and ecological services is the magnitude of those service flows.

The three stages in Figure 3.4 correspond to the three sets of relationships discussed above. Stage 1 involves an assessment of the current quantity and quality of the ground water resources in a particular area and an assessment of how events or circumstances might alter the baseline quantity and quality. This alteration could come about through underlying economic and social forces, such as increased population, or arise from some explicit decision, such as a change in management or policy or institutional structures. This stage represents a crucial input in estimating the economic value of the stock and makes explicit the role of natural sciences in the valuation process.

The second stage maps changes in ground water resources into changes in the service flows from the use of the resource. Stage 3 represents the formal economic analysis, equivalent to the third set of relationships. This third stage quantifies the value of services and how these values are affected by changes in service flows.

Each of the stages in Figure 3.4 requires in-depth research and is accompanied by its own levels of uncertainty. The uncertainty with respect to the estimates of the biophysical impacts on the quantity or quality of the resource will be carried through the valuation process and will be compounded by the uncertainties in the economic valuation methods. For example, the National Research Council's *Ground Water Vulnerability Assessment* (NRC, 1993b) attests to the importance as well as the difficulties and uncertainties present in current vulnerability assessment methods available to predict changes in the quality and quantity of ground water resources.

The conceptual framework involves the research of economists, building upon the hydrological and biophysical analyses that preceded it. The uncertainties and challenges associated with economic valuation techniques as they pertain to valuing ground water assets are discussed in Chapter 4. The hydrological, physical, and biological principles relevant to the economic valuation procedures were discussed in Chapter 2.

As Boyle and Bergstrom (1994) indicate, "Economic valuation of ground water therefore requires that progress be made on two fronts: establishing formal linkages between ground water policies and changes in the biophysical condition of ground water and developing these linkages in a manner that allows for the estimation of policy-relevant economic values." While each of the stages in Figure 3.4 can be associated with specific disciplines, one cannot overemphasize the need for interactions and cooperation among economists, other scientists, and water managers to value ground water resources.

Relationship to Benefit-Cost Analysis

The framework proposed in this chapter for valuing ground water could just as well be termed a framework for measuring the economic benefits of ground water. Information obtained from an analysis of the benefits of ground water would be used in a full fledged benefit-cost analysis (BCA) of regulatory actions or management decisions affecting ground water quantity and quality.

Benefit-cost analysis has had a long history relating to water resources. The U.S. Bureau of Reclamation and the U.S. Army Corps of Engineers initially developed BCA to evaluate surface water investments. The overall objective was to provide a picture of the costs and gains associated with investments in surface water development projects.

In more recent years, BCA has been applied to environmental and resource regulations. (For details see Kneese, 1984.) In these applications BCA should not be used as a simple decision rule but rather as a framework and a set of procedures to help organize available information and evaluate tradeoffs. Viewed in this way, the framework proposed in this chapter is an approach to quantifying the benefits of current and proposed management practices affecting ground water. If this information were to be used in a decision-making framework, it would need to be matched with information on the costs of alternative management strategies.

RECOMMENDATIONS

- **As noted earlier, some knowledge of a resource's TEV is vital to the work of water managers, and in the development of policies dealing with allocation of ground water and surface water resources. For many purposes, the full TEV need not be measured, but in all cases where a substantial portion of the TEV will be altered by a decision or policy, that portion should be measured.**
- **Policy-makers must recognize the role of the discount rate in ensur-**

ing the efficient allocation of resources over time. As such, the discount rate should reflect the opportunity cost of financing ground water projects. Issues of equity or fairness should be addressed directly and not through adjustments to the discount rate.

- **An interdisciplinary approach, such as the conceptual model presented in Chapter 3, is useful in conducting a ground water value assessment. The approach should incorporate knowledge from the economic, hydrologic, health and other social, biological, and physical sciences. Every assessment should be site specific and integrate information on water demands with information on recharge and other hydrologic concerns, and to the extent possible, should reflect the uncertainties in both the economic estimates of the demand for ground water and in the hydrologic and biophysical relationships.**
- **There are many research needs related to natural resource valuation concepts and methods. Research is needed to:**

a. determine the general circumstances under which nonuse values are likely to be significant;

b. provide a clearer understanding of how changes in water quality alter value; and,

c. develop better methodologies for linking ground water policy and changes in the biophysical properties of aquifers. Such research must be multidisciplinary.

REFERENCES

Boyle, K. J., and J. C. Bergstrom. 1994. A framework for measuring the economic benefits of ground water. Department of Agricultural and Resource Economics Staff Paper. Orono: University of Maine.

Freeman, A. M. III. 1993a. The Measurement of Environmental and Resource Values: Theory and Methods. Washington, D.C.: Resources for the Future.

Freeman, A. M. III. 1993b. Non-use values in natural resource damage assessments. Pp. 161-162 in Valuing Natural Assets, the Economics of Natural Resource Damage Assessment, Kopp and Smith, eds. Washington, D.C.: Resources for the Future.

Kneese, A. V. 1984. Measuring the Benefits of Clean Air and Water. Washington, D.C.: Resources for the Future.

Kopp, R. J., and V. K. Smith, eds. 1993. Valuing Natural Assets, The Economics of Natural Resources Damage Assessment: Washington, D.C.: Resources for the Future.

Lind, R. C. 1990. Reassessing the government's discount rate policy in light of new theory and data in a world with a high degree of capital mobility. Journal of Environmental Economics and Management 18(2):S8-S28.

National Research Council. 1993a. Sustaining Our Water Resources. Washington, D.C.: National Academy Press.

National Research Council. 1993b. Ground Water Vulnerability Assessment. Washington, D.C.: National Academy Press.

National Research Council. 1994. Alternatives for Ground Water Cleanup. Washington, D.C.: National Academy Press.

Page, T. 1977. Conservation and Economic Efficiency. Baltimore, Md.: Johns Hopkins University Press.

Page, T. 1988. Intergenerational equity and the social rate of discount. Pp. 71-89 In Environmental Resources and Applied Welfare Economics: Essays in Honour of John V. Krutilla, V. K. Smith, ed. Baltimore: Resources for the Future Press.

Smith, V. K., ed. 1988. Environmental Resources and Applied Welfare Economics: Essays in Honour of John V. Krutilla. Baltimore: Resources for the Future Press.

Tsur, Y., and T. Graham-Tomasi. 1991. The buffer value of ground water with stochastic surface water supplies. Journal of Environmental Economics and Management 21: 201-224.

Tsur, Y., and A. Zemel. 1995. Uncertainty and irreversibility in ground water resource management. Journal of Environmental Economics and Management 29(2):149.

4

Economic Valuation of Ground Water

Chapter 3 presented an integrative framework for valuing ground water resources. This chapter examines the key economic principles and methods used to value various ground water services identified in the previous chapter. It is divided into five major sections. The first offers a brief history of the science and art of economic valuation of natural/environmental resources, including the role of these methods in public policy development. This is followed by a review of the methods for estimating the value of environmental amenities. The approaches are discussed in terms of their relevance for the categories of service flows generated from the integrative framework in Chapter 3. The fourth section reviews selected ground water valuation studies with the aim of drawing conclusions about the state of current knowledge of the value of ground water resources. Finally, recommendations are made for using elements of the integrative framework from Chapter 3 and the economic concepts and methods presented here to estimate the value of ground water in specific contexts. The application of these methods to a range of ground water services is explored in a series of case studies in Chapter 6.

HISTORY OF ECONOMIC VALUATION OF NATURAL/ENVIRONMENTAL RESOURCES

Since the 1960s economists have developed a variety of techniques for assessing the value of nonmarket goods and services, not priced and traded in markets. While most applications are to natural resources and environmental assets, the concepts and methods of nonmarket valuation extend to a range of goods not usually traded in markets. The ability to assign values to such goods and services has improved the accuracy of benefit-cost analysis. Inclusion of economic values for some important (and previously ignored) classes of environmental services enables benefit-cost assessments to reflect more fully the consequences of natural resource policies and regulations.

Some of the earliest attempts to value a nonmarketed natural resource involved the value of water to agriculture in the western United States. Since water has traditionally been allocated to farmers and other users according to the prior appropriation doctrine ("first in time, first in use"), information was not available on the user's willingness to pay for water. To estimate (impute) a value for irrigation water, economists used models and techniques borrowed from studies of the behavior of firms, such as profit-maximizing models of farm behavior cast as linear or other programming models. Specifically, economists had to infer value by examining changes in returns to the farm associated with changes in the amount of water applied. In this way they could estimate the value of both surface and ground water.

These early water resource valuations used conceptual models and estimation techniques that had been developed and used primarily for analyzing market-related issues. These techniques worked well in assigning an economic value to water use in agriculture, given that water is simply an input into the farm's production process and that abundant cost data (on other inputs) and revenue information for farm operations existed.

The first application of techniques developed specifically for valuing nonmarketed commodities involved the travel cost method (TCM), Hotelling proposed in 1946 as a means of valuing visits to national parks. The travel cost method, in its numerous variants, has been used extensively to assess the value of a commodity used directly by the consumer, namely outdoor recreation. Refinements of the travel cost method and the development of new techniques, such as the contingent valuation method (CVM) and hedonic price method (HPM), enhanced the ability of economists to value a wider range of use values for environmental commodities, including improvements in air and water quality. Within the past decade, attention has shifted to estimating nonuse values, such as what individuals are willing to pay to ensure the existence of species or unique natural settings. The values elicited with these techniques for specific environmental goods and services are being used in an increasing array of settings; however, their use is not without controversy, as discussed later in this chapter.

The development of nonmarket valuation techniques enabled economists to place values on individual environmental commodities. However, policy and regulatory attention is now increasingly focused on the management of ecosystems. Valuing complex hydrologic or ecological functions and the associated range of service flows is relatively uncharted territory and raises a number of conceptual and practical issues. For instance, natural scientists cannot unambiguously define and measure ecosystem performance and endpoints. Other problems arise from the inability of economic science to measure adequately the consequences of long-term and complex phenomena. A related problem is the difference in disciplinary perspectives between economists and scientists from other fields who provide knowledge about physical relationships required for bioeconomic assessments, such as how a change in aquifer flow will alter surface stream flow and how a change in stream flow will, in turn, affect items people value, such as recreational fish catch. These issues and challenges affect the ability of economists to assess the full range of service flows from ground water; these challenges are discussed in the case studies in Chapter 6.

THE ECONOMIC APPROACH TO VALUATION

Economic values are only one type of assigned values (Brown, 1984). They indicate human preferences for a good or service and are not inherent in the good or service itself. Further, economic values are exchange values; they reflect the terms of trade, dollars for services. Decision criteria which are based on economic values, such as efficiency and benefit-cost analysis, demonstrate a utilitarian philosophical perspective. Recognizing and using economic values does not deny the existence or validity of alternative perspectives of value; however, the foundations of economic analysis offer the only unifying approach in making some types of private and public choices.

The Role of Time in Economic Valuation

Ground water services, like the services arising from many natural resources, frequently occur over multiple time periods. The rate of conversion of value between time periods is called a rate of time preference. The rate of time preference is defined at the individual level, and is a feature of people's desires. If an individual's rate of time preference is positive (greater than 0 percent), then the individual prefers a dollar today to a dollar a year from today because the dollar (or the consumption that dollar could purchase) in one year is worth less to the individual than the value of a dollar (and its level of consumption) today.

To account for this, some economists like to discount the future values of assets in order to compare them accurately to present assets. Discounting converts future values to present ones. The present value (V) is related to a future value (FV) received t years hence by the rule

$$(1) \quad V = FV/(1+r)^t$$

in which r is the role of time preference. Discounting thus reduces the future value of an asset by a percentage equal to the rate of time preference. Note that the two concepts of a rate of time preference and a bank rate of interest are distinct. They are, of course, related to one another in a market system. (Indeed, bank interest is an implicit recognition that people value a dollar more today than the same dollar tomorrow.)

The role of changes in productivity, as discussed in the following section, is also important in determining the appropriate discount rate. The following two examples demonstrate how the concepts of rate of time preference, discounting, and present value are used in measuring economic values over time. The examples include calculation of the value of an asset and the optimal rate of extraction of a resource over time. Both examples are relevant to the valuation of ground water services.

The Value of an Asset

An asset, such as a piece of machinery or a ground water aquifer, is valuable because of its contribution to producing a product of value (e.g., agricultural crops or clean drinking water). The relationship between the value of the product produced and the value of the machine or an aquifer is important. Suppose that a machine or an aquifer lasts forever and that it contributes an increment to production each year that the firm values at $\$R$. Suppose further that the bank rate of interest is i percent. Then value (V) of the asset is

$$(2) \quad V = R + R/(1+i) + R/(1+i)^2 + R/(1+i)^3 + \ldots.$$

$$(3) \quad = R/i.$$

The value of the asset today is thus equal to the sum of the annual incremental contributions the asset will make to production during its life, less an appropriately discounted percentage for each year. This is the value (V) of the machine or aquifer to the firm; and the firm would be willing to pay up to this amount (but no more) today for the asset. In short, the value of any productive asset is the present value of the increment to the owner's objectives that it will generate. The relationship in (3) holds exactly only for infinitely lived assets that do not depreciate, but the same idea holds in general. In this special case, we can see that the machine's value is such that the yearly increment to the value of production, R, (called the rental value of the machine, for that is what the company would be willing to pay to use the asset for one year) is the interest rate times the value of the asset.

The Dynamic Price of Water

The example above shows one way of placing a value on the services provided by a ground water aquifer that produces a finite stream of benefits. A somewhat more complex dynamic decision involves the optimal time rate of use (exploitation) of a natural resource. Optimizing involves balancing marginal gains against marginal costs. Suppose a single private firm owns an aquifer. For now, suppose further that the aquifer is confined, with no recharge. Thus it is a finite exhaustible resource, like a mineral deposit. The stock of water contained in the aquifer is known to be S (for stock) gallons initially. After t years of extraction, there are $S(t)$ units of water left in the aquifer. The firm extracts an amount E (an action corresponding to extraction) of water; in year t, this amount is $E(t)$. Suppose this extracted water can be sold for a price of $\$P$ per unit. The dollar cost of pumping and distribution depends on both the amount extracted and the size of the stock. A larger stock means lower pumping costs. To capture this idea, let $C(S)$ be the unit cost of pumping and distributing water when the stock size S gallons; total cost is $E(t)C(S(t))$.

The objective of a private water supply company is to maximize the present value of extraction. To do so, the firm will balance the benefits of an additional (marginal) unit of extraction against the (rising) costs of removal; that benefit will be P, the price the unit sells for. The marginal costs of extraction will be of three kinds. First there is the marginal pumping and distribution cost $C(S)$. Second, there is the opportunity cost of current extraction: that is, the loss of the option to extract that unit of water later. Third, pumping water today increases the cost of pumping at all future times. Thus there is a "dynamic" cost of pumping water that includes not just the usual cost of extraction and distribution but opportunity costs and the "cost" of driving up future pumping costs.

The dynamic cost of water increases as the ground water is depleted. Let $R(t)$ be the dynamic cost at year t. Balancing price and marginal extraction cost will involve accounting for both the unit cost of pumping ($C(S)$) and the dynamic cost ($R(t)$) as in

$$(4) \quad P = C(S) + R(t).$$

As extraction continues, $C(S)$ rises while S declines. In the market, the price of water will rise. The dynamic term $R(t)$ also increases over time to reflect increasing scarcity of water.

If there is recharge, the details of the model change, but not its fundamental lessons. There still is a dynamic price of water, $R(t)$, but its behavior over time is modified to reflect recharge. At some point the aquifer may enter a steady state, in which the amount of extraction and the amount of recharge are equal and no net change in the stock takes place. Then, assuming energy and other costs remain stable, the price of water becomes a constant as well, equal to the stable extraction and dynamic costs $C(S) + R$. It should be noted that in circumstances where aquifers discharge naturally to a stream, assuming that extraction does not affect future uses or users, and the level of the water table is unaffected—then ground water is not scarce and $R(t)$ equals zero.

The term $R(t)$, the dynamic cost of the additional water, is also the rental value of the ground water stock. It is the amount the firm would pay for another unit of ground water stock. As such, it measures the value in the market of having another unit of ground water in terms of the extra value the ground water will produce either as a consumption good or as an input to the production of other goods. This is the value the marketplace places on additional ground water resources. This may or may not correspond to the best thing for society, depending on society's objectives.

The dynamic price of water, $R(t)$, gives the value of having another unit of stock. It is also a price for balancing the (dynamic) supply of water against the demands for water, present and future. Obviously, its magnitude depends on several things. First, $R(t)$ depends on the stock of water. If all else is equal, as the stock goes up, $R(t)$ goes down and vice versa. In instances where ground water is not scarce, it commands no rental value and $R(t)$ is zero. What is relevant to proper water pricing in a market is the size of the stock relative to demand for it. Anything that increases the demand for the ground water stocks (e.g., population growth or increased allocation of water to produce environmental services) increases $R(t)$. And conversely, decreases in demand (by water conservation or development of substitute sources) will reduce the efficient water price.

Contamination events also will drive up the dynamic water price, $R(t)$ by reducing usable supply. This allows a method for determining the social cost of contamination. If contamination makes ground water useless for some purpose (e.g., drinking) but it leaves it acceptable for another (e.g., irrigation), the stock relative to the second demand will increase. This will automatically be built into changes in the dynamic price.

METHODS FOR ESTIMATING THE ECONOMIC VALUE OF NATURAL/ENVIRONMENTAL RESOURCES

The preceding section provided a brief introduction to economic concepts and constructs central to the measurement of benefits and costs. Applied economic analysis uses these theoretical concepts and constructs in combination with models and quantitative techniques, to answer questions involving private and public choice. Specifically, theories of firm and consumer behavior are used to develop testable hypotheses and as a guide to model specification. Quantitative methods, such as econometric and operations research techniques, provide a means of testing the hypotheses and models against real-world data. This combination of models and techniques has been successfully used for decades to address a wide range of economic issues, including the estimation of values for nonmarketed commodities.

The place to start any valuation effort is to look for situations where prices for natural/environmental resources are already revealed as a result of competitive market or simulated exchange arrangements (Freeman, 1993). Many natural resources are sold in markets and therefore the prices that result offer opportunities for valuing natural resources. These markets must be well-functioning and competitive in order for the prices to reveal reliable information. It should be remembered that prices represent only a marginal value and steps must be taken to calculate the total value by estimating the demand for the good (see Chapter 3). Nonuse values cannot be captured through this approach.

Recent developments in negotiated land transactions also offer an opportunity to gain some important information about the value of a natural resource. For example, some municipalities in the West bought up agricultural land in order to obtain water rights. These negotiated transactions over water rights provide evidence of the value of ground water in these areas.

Nonmarket valuation techniques consist of two basic types. Indirect approaches rely on observed behavior to infer values. Direct approaches use survey-based techniques to directly elicit preferences for nonmarket goods and services. Both sets of techniques share a foundation in welfare economics, where measures of willingness to pay (WTP) and willingness to accept (WTA) compensation are taken as basic data for individual benefits and costs.

Indirect Valuation Approaches

Indirect approaches, sometimes referred to as revealed preferences approaches, rely on observed behavior to infer values. This section begins with an overview of two general classes of indirect methods: derived demand and production cost techniques, which impute the value of a nonmarketed environmental input, such as ground water, into a production process; and the opportunity cost approach, which quantifies the economic losses associated with the impacts environmental degradation has on human health. The discussion then turns to more detailed presentations of three techniques that are commonly labeled as indirect methods: the averting behavior method, the hedonic price method, and the travel cost method. These methods depend upon the ability of individuals to discern changes in environmental quality and adjust their behavior in response to these changes. Recent summaries of indirect approaches can be found in Braden and Kolstad, 1991; Mendelsohn and Markstrom, 1988; Peterson et al., 1992; Smith, 1989, 1993; and Freeman, 1993. A summary of the advantages and disadvantages of the indirect as well as direct methods is given in Table 4.1.

Derived Demand/Production Cost Estimation Techniques

Where water is an important component of a production process and a firm's cost structure is known, the water's implicit value can be calculated by measuring water's contribution to the firm's profit. If water supply is unrestricted, a firm will continue to use units of water up to the point where the contribution to profit of the last unit is just equal to its cost to the firm. Even if water is "free," there will be costs to the firm associated with water use (including pumping and delivery costs). If water supply is restricted (for example, by quotas or water rights), the firm may cease use of water before the equality is met.

TABLE 4.1 Advantages and Disadvantages of Selected Valuation Methods

Method	Advantages	Disadvantages
Derived demand/production cost estimation techniques	Based on observable data from firms using water as an input or from household consumption. Firmly grounded in microeconomic theory. Relatively inexpensive.	Not possible to measure *in situ* or nonuse values. Understates WTP.
Cost-of-illness method	Relatively inexpensive.	Omits the disutility associated with illness. Understates WTP because it overlooks averting costs. Limited to assessment of the current situation.
Travel cost method (TCM)	Based on observable data from actual behavior and choices. Relatively inexpensive.	Need for easily observable behavior. Limited to resource use situations including travel. *Ex post* analysis; limited to assessment of the current situation. Does not measure nonuse values. Possible sample selection problems and other complications relate to estimate consumer surplus.
Averting behavior method	Based on observable data from actual behavior and choices. Relatively inexpensive. Provides a lower bound WTP if certain assumptions are met.	Estimates do not capture full losses from environmental degradation. Several key assumptions must be met to obtain reliable estimates. Need for easily observable behavior on averting behaviors or expenditures. *Ex post* analysis; limited to assessment of current situation. Does not estimate nonuse values.
Hedonic pricing method (HPM)	Based on observable and readily available data from actual behavior and choices.	Difficulty in detecting small, or insignificant, effects of environmental-quality factors on housing prices.

continued

TABLE 4.1 Continued

Method	Advantages	Disadvantages
Hedonic pricing method (HPM)		Connection between implicit prices and value measures is technically complex and sometimes empirically unobtainable.
Market prices or negotiated transactions	Based on observable data from actual choices in markets or other negotiated exchanges.	Does not provide total values (including non-use values) *ex post* in nature, limited to assessment of current situation. Potential for market distortions to bias values.
Contingent valuation method (CVM)	*Ex ante* technique: it can be used to measure the value of anything without need for observable behavior (data). Only method to measure existence or bequest values. Technique is not generally difficult to understand.	Since hypothetical, not actual, market transactions or decisions are the focus of CVM, various sources of errors (i.e., incentives to misrepresent values, implied value cues, and scenario misrepresentation) may be introduced. Expensive due to the need for thorough survey development and pretesting. Concerns about reliability for calculating nonuse values (particularly for such calculations to support natural resource damage assessments for use in litigation). Controversial, especially for nonuse value applications.

The level of water use at varying costs to the firm defines a "derived" demand relationship, given that the demand for the input (water) is derived from the demand for the output (e.g., agricultural commodities). Simple budgeting or more complex linear programming and other optimization methods have been applied to calculate use value and derived demand for ground water in agricultural production to gauge efficiency of water allocation or to manage ground water extraction rates (Snyder, 1954; Ciriacy-Wantrup, 1956; Burt, 1964, 1966; Bain et al., 1966; Kelso et al., 1973).

Production/cost techniques have also been applied to municipal water delivery and use (Teeples and Glyer, 1987). A related and important category of research on water values focuses on the *demand* for municipal water. Such studies do not use indirect techniques or processes to impute water value; rather, they combine concepts from the theory of consumer behavior with econometric (statistical) procedures to estimate the demand for water. This line of inquiry has documented consumers' willingness to pay for water under a range of prices and delivery systems (e.g., Wong, 1972; Berry and Bonen, 1974; Foster and Beattie, 1979; Cochrane and Cotton, 1985). These types of studies have also been helpful in understanding the "price-responsiveness" or price elasticity of water demand (Martin and Wilder, 1992; Renzetti, 1992). Application of these techniques to measure demand for (and value of) water requires sufficient variation in water prices across time and/or space to elicit statistically robust results. This condition is often lacking in municipal water pricing, where consumers or households often face a fixed price, regardless of quantity consumed.

Some of these input-oriented valuation techniques are conceptually similar to the averting behavior approach discussed in the next section, in that a lower bound on the value of water is indicated by what a firm spends to acquire water of acceptable quality. For agriculture, this expenditure may be for energy to pump ground water or for delivery systems to transport water to the site of use.

This general class of techniques can also be used to assess buffer value and other dynamic functions of an aquifer, such as the value of a ground water supply to supplement surface water during times of drought. Tsur and Graham-Tomasi (1991) used dynamic programming methods to estimate the buffer value of ground water to wheat growers in southern Israel's Negev region. Using certain assumptions, they found that buffer values were positive and in some scenarios were a significant component (up to 84 percent) of the total value of ground water. This application also highlighted the potential for uncertainty in surface water availability, acting through the buffer role of ground water, in influencing ground water extraction over time. This influence is a function of size of the aquifer stock, its extraction cost, and uncertainty. Moreover, differences in the magnitude of the buffer value of ground water have important implications for the dynamic behavior of ground water extraction (Tsur and Graham-Tomasi, 1991).

Using this class of static and dynamic optimization techniques requires detailed production and cost data. Such data are most likely to be associated with the production of marketed goods, such as agricultural production. Since the majority of potential ground water services do not fall into this use class (of inputs used in the production of marketed goods), the use of these techniques is restricted to some of the potentially less important ground water services.

Using Opportunity Costs to Value Health Losses (Cost-of-Illness Method)

Human health effects are a prime concern in ground water contamination incidents. Exposure to unsafe levels of substances in water through ingestion in drinking water or other routes (e.g., skin absorption) can lead to increased morbidity or mortality. In most cases contaminant levels are not high enough to produce acute health effects. Rather, consumption of relatively low levels of harmful substances in water may lead to long-term or chronic illnesses, such as cancer, and possibly to premature death. In addition to mortality losses, contamination of ground water creates losses due to increased morbidity, such as the costs of medical treatment and care, loss of leisure-time activities, and pain and suffering associated with illnesses (Spofford et al., 1989). The theory underlying WTP approaches to valuing mortality is summarized in Freeman (1993).

The two main approaches economists have used to value morbidity are based on either individual preferences (WTP or required compensation) or the resource or opportunity cost approach (Freeman, 1993). In the latter, known as the cost-of-illness (COI) approach, the analyst attempts to measure benefits of pollution reduction by estimating the possible savings in direct out-of-pocket expenses resulting from the illness (e.g., medicine, and doctor and hospital bills) and opportunity costs (e.g., lost earnings associated with the sickness). For example, the costs per illness or losses in wages per day associated with cancer caused by drinking water containing a volatile organic chemical would be multiplied by the number of days of illness in the population to arrive at an aggregate benefit figure.

The cost-of-illness approach has several important limitations. First, it does not consider the actual disutility of those afflicted with illnesses. Second, it overlooks that individuals faced with pollution undertake defensive or averting expenditures to protect themselves. Harrington and Portney (1987) demonstrated theoretically under a set of plausible assumptions that without the inclusion of expenditures on averting behaviors, the COI benefit estimation method will underestimate true willingness to pay for a reduction in pollution.

Averting Behavior Method

Actions taken to avoid or reduce damages from exposure to ground water contaminants are another category of economic losses. Theoretical explanations of averting expenditures are based on the household production function theory of consumer behavior. In the context of averting behavior models, the household produces consumption goods using various inputs, some of which are subject to degradation by pollution. The household may respond to increased degradation of these inputs in various ways that are generally referred to as averting or defensive behaviors.

The adverse impacts of ground water contaminants can be avoided in at least three ways: (1) buying durable goods (e.g., point-of-use treatment system); (2) buying nondurables (e.g., bottled water); and (3) changing daily routines to avoid exposure to the contaminant, such as (a) boiling water for cooking and drinking or (b) reducing frequency or length of showers if a volatile organic chemical were present (Dickie and Gerking, 1988). Households, businesses, and other organizations may undertake averting actions to protect individuals from exposure to contaminants.

Several theoretical analyses (Courant and Porter, 1981; Bartik, 1988) of the averting behavior methods have concluded that under certain conditions such expenditures can provide a lower bound estimate of the true cost of increased pollution. Averting expenditures and true benefits of a pollution reduction differ because such expenditures do not measure all the costs related to pollution that affect household utility. While this approach measures household production costs, it fails to capture direct utility losses related to pollution (Musser et al., 1992). Courant and Porter (1981) found that when the level of ambient environmental quality conditions is valued directly by individuals, it is uncertain whether averting expenditures are not necessarily an accurate lower bound estimate of pollution reduction benefits. Bartik (1988) concluded that theoretically correct measures of WTP can be estimated using averting expenditures for both marginal and nonmarginal pollution changes. The ability of this valuation approach to provide a lower bound to WTP depends on the following assumptions: averting inputs should not serve in the production for only one output that is valued by the household (i.e., no jointness in household production); households should not obtain direct utility from the averting behavior; no income effects occur as a result of loss of work through illness; and the purchases of durable goods do not lower costs. In many ground water contamination situations, at least one of these assumptions is not likely to hold. Care must be taken in interpreting averting expenditures alone as a lower bound estimate of the value of a ground water function or service. In most cases information from averting cost studies will need to be coupled with and in some cases compared to results from studies using other valuation techniques to arrive at a complete measure of value of the ground water (Abdalla, 1994).

Hedonic Price Method

The hedonic pricing method (HPM) is based on the premise that people value a good because of the attributes of that good rather than the good itself. For example, the decision to purchase a particular house may be influenced by the attributes of that house (number of bedrooms, square footage, view, quality of the neighborhood, etc.). If one of those attributes is an environmental commodity, such as clean air, comparison of the price consumers pay for houses in areas of "clean air" (all other attributes of the house being equal) may provide information on the value of clean air.

Hedonic price models encompass both land (housing) price models and wage models that account for variations due to environmental attributes (e.g., air and water quality, noise, aesthetics, and environmental hazards). Wage models can be used to infer values for environmental attributes by examining the relationship between wage rates and the quality of the environmental attributes across jobs and locations. Hedonic models can only measure use values. The measurement of use values is based on one fairly strong assumption (weak complementarity), which holds that the purchase of some market good is associated with consumption of an environmental good or service, and when consumption of the market good is zero, then demand for the environmental good or service is also zero (Adamowicz, 1991).

The hedonic technique, like other indirect nonmarket valuation methods, depends on observable data resulting from the actual behavior of individuals. An advantage of the HPM is that market data on property sales and associated characteristics are readily available from county or municipal sources (e.g., assessor's office) as well as from private real estate services. These data can usually be linked to other secondary sources of data for the same geographical area (e.g., data on water quality, air quality, or a range of physical attributes). These secondary sources of data can be used to construct indices of environmental quality for use in a statistical analysis.

Despite the advantage of readily available data, several problems limit the use of the HPM in many settings. One problem is that the effect of an environmental attribute or characteristic on price may be small and hard to detect statistically or to disentangle from the effects of all other variables. Another problem with the technique is that it is difficult to derive value measures from the estimated hedonic price function (the basic first-stage equation where the sale price of a house is regressed on the set of attributes of that house). Derivation of the value of an attribute requires a second-stage procedure to obtain a demand or WTP function built around market segmentation (to address an estimation problem known as identification). To date, few empirical studies have successfully completed the second stage. Thus most studies report only the results from the hedonic price function, which gives an estimate of the marginal effect of an environmental variable on price.

A brief hypothetical example illustrates the use of an indirect approach to measuring nonmarket value. The HPM can be applied to housing prices to estimate the value of environmental attributes, such as well (drinking) water or proximity to wetlands, which vary across a region. It is assumed that variations in housing prices can be linked to real or perceived variations in these environmental attributes (controlling for a variety of other statistical determinants). In practice the approach involves collection of cross-sectional data on house sales (or possibly assessed values) and information on a menu of potential determinants of value (lot size, number of bedrooms, etc.). These factors would include one or more indices of environmental attributes or services. Through multivariate statistical techniques, analysts can infer the marginal value of either positive or negative environmental externalities. For example, a researcher might find that the average homeowner in a particular county would pay \$*X* to be *Y* yards closer to an open-water wetland and would require a reduction in price of \$*Z* to purchase a house in an area of contaminated ground water.

There have been few hedonic price studies on ground water contamination problems and the results have not been conclusive. While they have found statistical differences for industrial sites (due to cleanup costs and liability concerns), researchers have been less successful in isolating the effect of contaminated ground water upon residential property values (Malone and Barrows, 1990; Page and Rabinowitz, 1993).

Travel Cost Method

Travel cost methods (TCM) encompass a variety of models, ranging from the simple single-site travel cost model to regional and generalized models that incorporate quality indices and account for substitution across sites. The basic premise behind all versions of the travel cost model is that the travel costs incurred in traveling to a site can be regarded as the price of access to the site. Changes in the travel cost to a site can then be viewed as having the same effect on visits to the site as would a change in an access fee or a price. Under a set of assumptions involving opportunity cost of travel time, purpose of the trip, availability of substitute site, and time spent at the site, it is possible to derive the individuals' demand for visits to a site as a function of the price of admission using the simple or basic travel cost model.

In many situations the analyst is interested in understanding the effect of substitute sites on demand for a given site or the effect of changes in quality of certain site attributes on visits to the site. These types of concerns can be addressed using more complex versions of the travel cost model. For example, the role of substitute sites on visitation can be addressed with multiple-site travel cost models or with discrete-choice travel cost models. Changes in site quality, such as improvements in water quality, fish catch, and so forth, can be estimated using the generalized travel cost model, the hedonic travel cost model, or similar specifications. Because of the flexibility of travel cost methods and the relative ease of collecting data necessary for estimation, researchers have relied extensively on these methods in deriving use values for a wide range of recreational activities.

Since most extractive uses of ground water do not involve travel, TCM has limited applications to valuing these uses. However, ground water can provide *in situ* services, such as recharging surface water and wetlands and dilution of contaminants that may support recreational services. Since many ground water aquifers are a source of recharge into surface water and wetlands, ground water may support a number of recreational services. The TCM could conceivably be employed to value such services, although its use may be limited because of the difficulty of determining the share of recreational value attributable to ground water.

Direct Valuation Approaches

Direct approaches to nonmarket valuation use survey-based techniques to directly elicit preferences. The hypothetical nature of these experiments requires that markets be "constructed" to convey a set of changes to be valued. While there are a number of variants on these constructed markets, the most common is the contingent valuation method (CVM). CVM is a survey-based procedure designed to elicit a respondent's WTP or WTA for an environmental change (see Appendix B).

A method related to CVM is conjoint analysis, which includes contingent ranking of behavior. Conjoint analysis refers to a general approach marketing researchers employ to predict behavior based on studies of consumers, using contingent comparisons of product attributes, including price. Federal decision-makers concerned with valuation issues as well as environmental economists are giving greater attention to conjoint analysis. For example, a recent statement by the National Oceanic and Atmospheric Administration (NOAA) recommends the use of conjoint analysis, indicating that attributes may be valued in terms of price and if replacement costs were used would provide decision-makers the ability to compare alternative service flows (NOAA, 1995). Environmental economists are exploring several variants of this approach, including contingent ranking and contingent behavior, as a way to improve upon the CVM. These survey-based techniques derive information about an individual's preferences between alternatives with varying levels of environmental attributes. The contingent ranking method goes beyond the simple yes/no of a referendum format and asks individuals to reveal more information about their preferences by asking them to rank the hypothetical alternatives. If one attribute of the good is measured in monetary terms, subsequent statistical analysis allows the calculation of the WTP for changes in the attribute. A disadvantage of contingent ranking is that it is time-consuming and potentially difficult for a respondent to rank several goods and multiple attributes.

The contingent behavior (or activity) method involves the use of hypothetical questions about activities related to environmental goods or services. The main use of contingent behavior surveys has been to support other valuation analysis. For example, it can be used to support a TCM study of benefits by assessing how participation in recreational activity changes as environmental quality changes. Such contingent choice information can then be used to estimate a shift in the demand curve for recreational visits (Freeman, 1993).

An Introduction to the Contingent Valuation Method

The contingent valuation method (CVM) can be viewed as a highly structured conversation (Smith, 1993) that provides respondents with background information concerning the available choices of specific increments or decrements in environmental goods. Values are elicited directly in the form of statements of maximum WTP or minimum WTA compensation for hypothetical changes in environmental goods. Typically, multivariate statistical techniques are used to model a WTP function. Such models allow the analyst to control for variation in the personal characteristics of the respondents, check for consistency of results with economic theory, and possibly estimate an entire WTP response across varying levels of environmental goods.

The contingent valuation method is applied when calculating for both use and nonuse values. The flexibility it provides in constructing hypothetical markets accounts for much of the technique's popularity. There are numerous methodological issues associated with application of CVM including how the hypothetical environmental change is to be specified, how valuation questions are formulated, the appropriate welfare measure to be elicited (i.e., WTP or WTA), and various types of response biases. Randall (1991) argues that because of the importance of nonuse values, CVM is likely to be the primary tool for measuring the environmental benefits of biodiversity. The CVM is also capable of measuring the disutility associated with some types of environmental degradation that indirect methods are unable to capture. Recent summaries of CVM can be found in Mitchell and Carson, 1989; Carson, 1991; Portney, 1994; Hanemann, 1994; and Diamond and Hausman, 1994.

CVM for Estimating Use and Nonuse Values

The contingent valuation method is a direct valuation technique: researchers ask people about their willingness to make certain trades and use the answers to estimate willingness to pay. Its appeal is that it is the only method that (in principle) can be used to estimate nonuse values for goods that do not yet exist or quality changes for goods that are outside the bounds of experience. It can also help estimate use values for goods traded on markets (in which it is called marketing research) or nonmarket goods such as ground water quality or recreation sites.

Three important classes of errors can bias the results of CVM studies (Freeman, 1993). First, participants may have an incentive to misrepresent their value for the hypothetical environmental good or service. For example, some respondents may state low values in order to reduce their obligation to pay for the goods or service even if they value it (i.e., free-rider behavior). Others may overstate the actual value if they believe the bid will affect the level of provision and the good is desired. A second category is implied value cues. This bias is particularly troublesome for unfamiliar goods or ones for which the respondent has not yet developed clear preferences. In such cases, the respondent may look for clues regarding a "correct" choice or value from the information provided by the researcher. When such "value cues" are present, they are likely to systematically bias the values elicited. One type of this problem is known as "starting point" bias. However, this bias can be overcome by using a referendum or voting approach to the bid question. Another form of bias, called "yea saying," may even exist in the voting format (Mitchell and Carson, 1989). A third category of possible error in CVM studies comes from mispecification of the scenario. This causes the respondent to have a different definition of the environmental good or service than the researcher intended.

Researchers have employed many versions of the CVM, but CVM practitioners now agree on certain best practices. The basic idea is to have an individual vote yes or no on a public program which provides a change in the provision of some good, such as air quality, and that will cost households like theirs $\$X$. The amount X is then varied across the sample. If a person votes yes, then WTP>X, while if they vote no, then WTP<X. Statistical techniques are then used to uncover willingness to pay. Sometimes, a follow-up question is asked: "If you vote yes at X, how would you vote at $X+c$, or if you vote no at X, how would you vote at $X-c$?" These "double-bounded" estimators provide more information on WTP.

This voting approach is called a dichotomous choice format. Most CVM studies use some form or adaptation of an open-ended question such as: "How much would you be willing to pay in increased fees, taxes, or prices, for q?" The voting approach has some desirable properties. First, it gives incentives to tell the truth. Second, people are familiar with voting on public programs, at least in many places: we do not purchase education; we vote on property tax increases. A disadvantage of this format is that it makes inefficient use of a sample and thus increases cost. However, this disadvantage can be at least partly overcome by asking respondents follow-up questions (Freeman, 1993).

A sound CVM study requires careful attention to development of the questionnaire (see Appendix B). Advances in psychology have enabled researchers to recognize circumstances in which individuals who say they intend to do something (e.g., vote yes on a public program) actually will do so. This intended-actual behavior link is what CVM attempts to establish: if respondents say they would pay $\$t$ for the program and were actually faced with the choice, they really would pay.

In addition, people must understand *exactly* the good on which they are voting and that they are voting only on that good. This can be difficult, since people are aware that we have only a limited understanding of how elements of the environment are interconnected. Thus statements that something *does not now and never will* do something else may not be plausible. Further, care must be taken that one program is not symbolic of a larger, implied program. Second, respondents must have some confidence that the program will actually supply the good. They should be considering the change in environmental quality, not whether there is some better means to provide that change than the proposed program. And third, respondents must search their preferences, taking the matter seriously and comparing a payment for this good to other things they can do with their money: both other public programs (education, crime fighting, etc.) as well as their own consumption.

To ensure these things, researchers must conduct qualitative research via focus groups and one-on-one interviews. Researchers need to make sure that the language they used in the final survey conveys exactly what they intend, and this can be a formidable task. Therefore, investigators must spend time with people, talking through "What were you thinking when I asked why?" and uncovering the impact of alternative approaches.

When the survey deals with past events, such as ground water contamination, the researcher must decide what type of program to present. One possibility is a hypothetical prevention program that would have protected the ground water if it had been in place before the contamination; another is an accelerated recovery/restoration program. The former is more what the investigator would like to sell, since it captures the whole event, but the latter may be more believable and easier to describe.

Much more can be said about CVM studies of particular types of issues, but such a detailed review is beyond the scope of this chapter. A large amount of research has been done on CVM over the past 25 years, and our understanding of it has expanded dramatically. It is clear that many CVM studies have produced meaningless WTP estimates and that adding a CVM question to the end of a telephone or mail survey without benefit of qualitative research to test the question is bad practice. It also seems clear that careful CVM research can generate reliable results, at least for some types of goods and values (e.g., use values).

While most economists accept CVM for direct use values, its application to measure nonuse values has been very controversial. Exactly how far CVM reliability can be extended to encompass unfamiliar goods and nonuse values has become the key issue. The feasibility of using CVM to measure some types of ground water services therefore remains in question.

A Special Problem: Estimation of Nonuse Values

Nonuse values are the most difficult to measure of TEV components. The contingent valuation method is the only technique available for assessing these values. The topic of existence values for environmental assets is one of the most controversial in environmental economics (Bishop and Welsh, 1992; Edwards, 1988; Kopp, 1992; Rosenthal and Nelson, 1992; McFadden, 1994; Hausman, 1993).

Examples of some of the ambiguities in existence value estimation can be seen in a CVM study on bald eagles, wild turkeys, and Atlantic salmon in New England. While Stevens (1991) found substantial economic benefit from protection and restoration programs, the results also indicated that in a setting of potential irreversibility, existence values were difficult to quantify and sensitive to how the species were aggregated. Further, a majority of respondents viewed species protection as important but were unwilling to pay for such programs. Follow-up questions indicated that many respondents were uncertain of their values or protected the WTP question for ethical reasons.

Much of the recent controversy over CVM and its use in eliciting nonuse values has been stimulated by questions surrounding natural resource damage assessment (NRDA) and liability cases. Sparked by the government's use of CVM in the Exxon Valdez oil spill case, the debate has focused on whether CVM can provide plausible estimates of value for individuals who may not be familiar with the good in question (i.e., individuals whose total value is made up entirely or largely of nonuse values). In 1992 NOAA, part of the U.S. Department of Commerce, convened a panel of blue-ribbon economists to provide guidance concerning the potential use of CVM in measuring lost nonuse values in promulgating regulations, pursuant to the Oil Pollution Control Act of 1990. The NOAA panel essentially reaffirmed application of CVM, provided rigorous guidelines are followed (Arrow et al., 1993). The panel recommended high-quality survey research (e.g., appropriate sampling and thorough pre-testing of instruments, etc.) and concentrating on more specific concerns related to CVM. The overall effect of the NOAA panel report is to make CVM very expensive and limit its application in many settings. A litigation-quality study conducted by a consulting firm for an NRDA in accord with the NOAA guidelines could cost several million dollars. Perhaps because of the increased cost of CVM studies and the continuing controversy surrounding the theoretical basis of CVM-based measurement of nonuse values, NOAA proposed new rules for assessing natural resource damages under the Oil Pollution Control Act of 1990 (NOAA, 1995). The new proposed rules eliminate "compensable values" in natural resource damage claims and instead focus on actions to restore natural resource services. The proposed rules thus downplay valuation of resources (including nonuse values). Values, including those from CVM studies, may still be used in making restoration decisions.

Not all CVM studies need be done with the exacting care required for NRDA litigation. One of the open questions in this area is the reliability of less expensive CVM studies (done via mail rather than by in-person surveys, for example) regarding goods familiar to people (like water availability).

CURRENT KNOWLEDGE OF GROUND WATER VALUES

Chapters 2 and 3 of this report discussed the interdisciplinary nature of the ground water valuation process. Each discipline has made significant progress in understanding and modeling components of this valuation process. To obtain an accurate accounting of the value of ground water resources, we must combine these components into an assessment framework. This in turn means that each discipline must understand what information the other disciplines need in the assessment process.

For the economic component of the assessment framework, we need reliable and valid estimates of the benefits to society of ground water protection. This requires correct application of economic valuation techniques. This section reviews selected studies of ground water valuation, with emphasis on two categories of studies: those based on CVM and those that use averting behavior approaches. At the outset, we should note that past ground water valuation studies have focused primarily on a small part of the known ground water functions and services identified in Chapter 3. Thus our current empirical knowledge of the values of ground water is quite limited.

Results of Indirect Approaches

Relatively few empirical studies of ground water values have been conducted employing indirect methods. Of the studies that focus on services related to ground water quality, the averting behavior approach has been most commonly used.

Ground Water Studies Using the Averting Behavior Method

At least five studies have used the averting behavior approach to measure household-level costs associated with ground water contamination. As noted earlier, depending upon whether key assumptions are met, the results of such studies may not accurately represent lower bound estimates of WTP for ground water services. Also, values obtained from averting behavior methods must be combined with estimates of other ground water services to get an estimate of the total value of ground water. Despite these limits, results from carefully done averting behavior studies can provide important information needed for policy-making. For example, as a lower bound estimate of benefits of ground water protection, they can be used as an initial screening step in comparing benefits and costs of protection alternatives and in helping to decide if more in-depth valuation efforts are needed (Abdalla, 1994). The results of five averting behavior studies are highlighted below. Additional information on the studies can be found in Table 4.2.

Smith and Desvouges (1986) found in a sample taken in the Boston area that bottled water and water filters were purchased for the sole purpose of avoiding hazardous waste by 30 and 7 percent of households, respectively. Losses due to water quality degradation were not estimated, however, since they lacked detailed data on household averting behaviors and their costs.

Abdalla (1990) and Abdalla, Roach, and Epp (1992) documented averting expenditures of households served by public water systems in two Pennsylvania communities that had organic chemicals in their water supplies. At a central Pennsylvania site, 96 percent of the households were aware of water contamination and 76 percent of those with such knowledge undertook averting behaviors.

TABLE 4.2 Summary Information on Averting Behavior Studies

Author(s) Publication Date(s)	Study Location	Current Ground Water Condition	Type of Contamination
Smith and Desvouges, 1986	Suburban Boston	Uncertain—one town had experienced several prior episodes of contamination of drinking water by hazardous wastes	Hazardous waste
Abdalla, 1990	Township in Centre County, Pennsylvania	Contaminant in water for 26 weeks before new source provided. 96% knew of drinking water contamination	Perchloroethylene No drinking water standard in effect at time of contamination
Abdalla, Roach, and Epp, 1992; Roach, 1990	Borough in Bucks County, Pennsylvania	Contaminant in water for 88 weeks—43% knew of drinking water contamination	Trichlorethylene Drinking water standard exceeded
Powell, 1991	Selected communities in New York, Pennsylvania, and Massachusetts	7 Communities experienced contamination in past 10 years. 16% of households knew of contamination within last 10 years	Trichlorethylene in 6 communities; diesel fuel in one community (NY)
Collins and Steinbeck, 1993	West Virginia (statewide)	Private individual well owners with contamination problems that agree to survey	Bacteria, minerals, organics

[a]Percent of households taking averting actions of those aware of contamination

Source of Drinking Water	Avoidance Actions[a]	Average Annual Household Avoidance Cost	Average Annual Household Bottled Water Purchases
Public water supply	Bottled water Water filters	NA	NA
Public water supply serving 1,600 households	New bottled water purchase (47.8%) Increased bottled water purchase (15.2%) Boiling water (23.0%) Hauling water (29.3%) Water filter (3.3%)	$252 (1987)	$142 (1987)
Public water supply serving 2,760 households	New (11.1%) Increased (19.1%) Boiling water (27.8%) Hauling water (18.9%) Water filter (10.4%)	$123 (1989)	$75 (1989)
18% private wells; 82% public water supplies	Restricted water use (31%) Boiled water (26%) Bottled water purchase (17.5%) Supply cutoff (6.3%)	NA	$32 (1990)
Private individual water systems. 90% used ground water as source	Clean/repair water systems (56%) Water treatment, new source, contaminant source contol (45%)	$320 to 1090 (1990) depending on contaminant type	NA

Only 43 percent of the households in the southeast Pennsylvania site were aware of contamination. Of those, 44 percent undertook avoidance actions. Costs averaged $252 and $123 for each household that chose to avoid the contaminant in the central and southeast study sites, respectively.

Powell (1991) documented household bottled water expenditures as part of a CVM study of ground water benefits in eight "clean" and seven "contaminated" communities in Massachusetts, New York, and Pennsylvania. Even though almost half of the communities had recent contamination problems, only 16 percent of mail survey respondents indicated that their water had been contaminated. For those that were aware, the average household bottled water expenditure was $32 per year, about three times that spent in uncontaminated areas. Respondents aware of contamination were willing to pay $82 per year for increased water supply protection compared to $56 for those that were not. Households relying on private wells were willing to pay $14 per year more for protection than those served by public systems.

Collins and Steinback (1993) documented responses to knowledge of water contamination of rural households relying on individual wells in West Virginia. Eighty-five percent of those who were informed about their household's contamination problem were found to engage in averting activities. The most frequent actions were cleaning and repairing water systems, hauling water, and treatment. Information from mail and telephone surveys was used to compute a weighted average annual economic avoidance cost of $320, $357, and $1090 for households with bacteria, minerals, and organic contamination problems, respectively.

Direct Methods: CVM Studies of Ground Water Values

CVM, given its potential ability to measure all components of economic value, has been used in a number of studies to estimate ground water protection benefits. Boyle (1994) compared eight CVM-based ground water valuation studies as part of a review by EPA's Science Advisory Board of an EPA-funded study of the national level benefits of cleaning up ground water contaminated by leaching from landfills. He compared the results of a CVM study completed in 1992 by researchers at the University of Colorado (McClelland et al., 1992) on the national level benefits of cleaning up ground water degraded by landfill leachate to seven other quite diverse ground water valuation studies conducted using CVM. The national-level McClelland et al. study results were compared with estimates from state and community-level studies in Massachusetts (Edwards, 1988); Michigan (Caudill, 1992; Caudill and Hoehn, 1992); Georgia (Jordan and Elnagheeb, 1993); Wisconsin (Poe, 1993; Poe and Bishop, 1992); New York, Pennsylvania, and Massachusetts (Powell, 1991; and Powell and Allee, undated); New Hampshire (Shultz, 1989; Shultz and Lindsay, 1990; Shultz and Luloff, 1990) and Georgia (Sun, 1990; Sun et al., 1992). A more recent review of these CVM-based estimates is provided in Crutchfield et al. (1995). These authors also

TABLE 4.3 Summary of CVM Studies—Major Characteristics

Author(s), Dates Study Site	Contaminant	Value	Focus
McClelland et al., (1992) national sample	unspecified	option price WTP to remediate contamination from landfills	NA
Caudill (1992) Caudill and Hoehn (1992) Michigan	unspecified	option price WTP to reduce the probability of contamination	NA
Doyle (1991)	unspecified	WTP to remediate contamination	$114-$163/HH/yr
Edwards (1988) Falmouth, Woods Hole, Mass.	nitrates	option price WTP to reduce the probability of contamination	$815/HH/yr for 25% reduction in risk
Jordan and Elnagheeb (1993) Georgia	nitrates	option price WTP to reduce nitrate contamina-tion to safe levels	medians public $65.88/HH/yr private $88.56/HH/yr means public $120.84/HH/yr private $148.56/HH/yr
Poe (1993) Poe and Bishop (1992) Portage County, Wisconsin	nitrates	option price WTP to prevent nitrate contamination	NA
Powell (1991) Powell and Allee (undated) 15 communities in N.Y., Pa., Mass.	TCE in 6 communities diesel fuel in 1	option price current value of respondents subjec-tive perceptions of safety	mean annual WTP $81.31 Mass $42.19 PA
Schultz (1989) Schultz and Lindsay (1990) Schultz and Luloff (1990) Dover, New Hampshire	unspecified	option price WTP protect/ maintain ground water quality	$40/HH/yr median
Sun (1990) Sun and Dorfman (1992) Dougherty County, Georgia	nitrates and pesticides	option price	means log model $998/HH/yr linear model $930/HH/yr empirical model $961/ HH/yr
Clemons et al. (1995) Martinsburg, West Virginia	unspecified	WTP for a Well-head Protection Program	median for nitrates $21.20/HH/yr median for VOC $13.68/HH/yr

use selected estimates from this literature to derive potential benefits of ground water quality protection in four areas of the United States.

Table 4.3 provides a general summary of some CVM ground water valuation studies completed to date. The table is adapted from Boyle and Bergstrom (1994) plus others studies cited above. The table, along with the following sections, summarize what has been added to the body of knowledge of benefits estimation using CVM. The next few pages focus on three important themes: 1) comprehensiveness of ground water functions and services; 2) effects of information on valuation responses, essentially the commodity definition problem; and 3) the ways that the studies have dealt with uncertainty.

Comprehensiveness of Ground Water Functions and Services

Measurement of the total value of the benefits of ground water protection requires a succinct definition of the services that ground water offers. These services include extractive services and *in situ* services (both use and non-use values). While it is generally recognized that measuring use values can be accomplished with several methods, the challenge of measuring nonuse values for nonmarket goods is daunting. CVM offers flexibility and the ability to measure nonuse values of public goods whereas other nonmarket valuation techniques such as the travel cost and hedonic pricing models are intended to measure use values (Powell, 1991). Most studies to date have focused on the health effects of ground water contamination, without including some other important functions from which humans derive value.

Most studies presented information on the health effects of ground water contamination (e.g., McClelland et al., 1992; Jordan and Elnagheeb, 1993; Poe, 1993; Sun, 1990). The others purposefully omitted references to the effect of contamination on human health by either telling respondents that their water was being monitored (Edwards, 1988) or eliciting respondents' perceptions of the safety of drinking. There is significant uncertainty as to the susceptibility of human health to ground water contamination, compounded by the difficulties of measuring dispersion of contaminants in ground water. Nitrate contamination provides a good example. The EPA has established upper limits for nitrate in drinking water because nitrates in drinking water are linked to two health problems, methemoglobinemia and gastric cancer. The incidence of illness and possibly death for the former is extremely low, and recent scientific evidence shows that the link to the latter is not statistically significant. Isolating the dose response effects of nitrates in drinking water is further complicated by the fact that humans also consume nitrates from other sources (e.g., preserved meat products).

All CVM studies to date have attempted to measure option price relative to the good's future availability in some cases and to its safety from the human health perspective. Option price is the maximum sum an individual would be willing to pay to change from their present level of risk to one of no risk (Freeman, 1993). As noted above, the human health focus ignores other functions of ground water that humans might value such as the use of ground water as a buffer stock, its role in other ecological functions, and its importance to economic development and agriculture. The exclusion of these other services and values may be a function of the current state of knowledge. Few studies have attempted to measure the value that people place on the ecological services that ground water supplies. Some of these functions have been addressed only in the physical sciences and measurement is complex. The time lag between a contamination event and its effects on ground water and the services it provides varies substantially according to physical characteristics of aquifers as well as the nature of the contaminant (Kim et al., 1993; Fleming et al., 1995). Moreover, the interplay between site specific characteristics, ecological functions, and resultant service flows makes possibility of benefits transfers difficult to assess.

Effects of Information on Valuation Responses

Commodity definition in CVM surveys often involves striking a compromise between a definition that is understandable and one that is technically accurate (Bishop and Heberlein, 1990). However, accurate information and definitions are essential in deriving accurate measures. When evaluating unfamiliar commodities, the less well-defined the commodity is, the greater the potential measurement error. Indeed, the likelihood that a respondent can even come up with a value decreases with lack of clarity of the commodity.

> In general, the descriptions of the current, reference, and subsequent ground water conditions are quite vague in the eight studies. This vagueness makes it difficult to establish the linkages between changes in ground water policies, ground water conditions, services provided, and estimated values. Of particular concern is the difficulty of ascertaining how the value estimates correspond to actual biophysical changes in ground water resources and the resulting service flows. (Boyle and Bergstrom, 1994)

Resolving this vagueness is problematic given the uncertainty surrounding measurement of ground water resources and the effects of contamination levels on human health. Reducing the uncertainty of actual biophysical changes may not be possible given the current state of knowledge. Also, it may run contrary to how individuals make decisions. For example, the individual may not understand contamination in parts per billion, but instead makes decisions about willingness to pay based upon subjective perceptions. Knowing that large portions of the water supply do not meet federal safety requirements would affect WTP. Knowing what those requirements are does not necessarily affect preferences, just as knowing that crossing a highway is dangerous does not require precise knowledge of the physics involved when a car meets a human or the probabilities of serious injury. The problem with subjective perceptions is that the baseline reference condition changes for each respondent; grouping respondents into a range of safety categories does not solve the problem since the difference between "somewhat safe" and "safe" is not the same for each respondent. Hence a change in condition from the initial to an improved environmental state cannot be compared nor aggregated for respondents.

There is some disagreement on what information should be presented in the hypothetical market (Boyle, 1994; Lazo et al., 1992). Specifically, what types of information, the quantity of information, and an appropriate presentation of technical information must be determined so that respondents provide valid and reliable valuation responses. Boyle (1994) recommends a hybrid of expert and respondent's subjective perceptions. The problem with this approach is that it is still not certain whether one or the other should be the starting point, which creates a lack of clarity regarding the effects of the different approaches on estimated values.

The baseline ground water condition is the foundation for determining values. Since many of the studies focus on option price (McClelland et al., 1992; Caudill, 1992; Edwards, 1988; Jordan and Elnagheeb, 1993; Poe, 1993; Poe and Bishop, 1992; Powell, 1991; Shultz, 1989; Sun, 1990; Sun et al., 1992), it is imperative that the baseline condition is well defined so that researchers can determine the welfare change from the initial condition to the proposed change. If the initial condition is not well defined, then it is questionable whether researchers are measuring what they intend to measure, and the validity of the results are called into question.

There are two general schools of thought regarding the presumed knowledge of respondents. The first (McClelland et al., 1992; Poe and Bishop, 1992) maintains that experts should be used to provide background information for survey design. This approach generates more consistent estimates, but suffers from testing bias and often results in informing respondents as to what they should answer. This information bias decreases the validity of results by making it difficult to generalize results from the informed sample to the general population. A second approach (Caudill, 1992; Edwards, 1988; Powell, 1991; Shultz, 1989) rests on the belief that consumers make decisions based on the information they have on hand, that is, their subjective perceptions of ground water characteristics (specifically safety). This approach may be more appealing from the standpoint that the validity of responses may increase but the random nature of responses about goods with which consumers have very little experience increases. Clemons et al. (1995) show that information on nitrates does not significantly alter value estimates, a finding contrary to those of Bergstrom and Dorfman (1994) and Poe and Bishop (1992).

Boyle (1994) recommends testing experts' opinions in focus groups with sample respondents to filter out highly technical information. Poe (1993) took steps to this end with a two-tiered study that provided respondents with water testing kits in the first stage so as to nail down the baseline condition. This approach allows a more reliable and valid measurement of the change in ground water condition from the initial condition to the new condition resulting from a ground water protection program because the initial condition is defined precisely through self-administered tests of well water quality.

Dealing with Uncertainty

The two preceding themes suggest that uncertainty is a common feature of existing CVM studies of ground water protection benefits. In measuring economic welfare under conditions of uncertainty, several factors must be considered when evaluating benefits estimates: future prices, future income, opportunity costs, uncertainty about future human health responses to prevent exposures, future use, and future availability.

Uncertainty about future states is compounded by any inability to measure present states. The studies reviewed here attempted to derive the option price or the maximum amount an individual would be willing to pay to maintain the option to consume the good. The conceptual model underlying the treatment of uncertainty uses the measurement of option prices for risk changes. Caudill (1992), Poe (1993), and Sun (1990) measured respondents' subjective perceptions of uncertainty of future supply, while Edwards (1988) found option prices for a range of probabilities and Powell (1991) asked respondents about their subjective perceptions of safety. Few studies have attempted to capture quasi-option value or the measurement of option price when there is a possibility of having better information in the future.

What We Know About Ground Water Values Based on Existing CVM Studies

The major issues of agreement and disagreement found in our review of these diverse CVM studies are presented in the next section. The framework established by Boyle and Bergstrom (1994) is a helpful guide to this discussion. Useful discussions of the strengths and weaknesses of CVM studies completed to date can be found in Boyle and Bergstrom (1994) and in Crutchfield et al. (1995).

Areas of Agreement (Strengths)

The CVM's ability to measure use values is generally accepted in the economics profession. Its ability to capture nonuse values remains controversial, even though the NOAA panel defined conditions under which CVM may generate reliable estimates of such values (e.g., adequate survey design and commodity definition). Efforts towards a consensus of survey design incorporating the use of verbal protocol and focus groups have led to the acceptance of CVM estimates in some policy settings (but not necessarily in litigation or judicial settings). There is also significant agreement that local context factors are important (e.g., Poe, 1993; Powell, 1991; Sun et al., 1992). For example, site-specific information is important to respondents when faced with a contingent decision.

Areas of Disagreement (Weaknesses and Areas for Future Research)

There is still some disagreement over payment vehicles, although most studies have focused on referendum type questions. Mitchell and Carson (1989) maintain that the chosen payment vehicle must be both realistic and neutral. Most studies have focused on either a referendum format (to pay for a bond for some type of protection or remediation program) or an increase in water or tax bills. Use of the latter presents difficulties because in instances where the respondent does not own property the vehicle is not realistic. Similarly, neutrality is problematic because a tax increase may invite scenario rejection. A referendum valuation question asks whether respondents would vote for the referendum given a specified cost for the referendum.

The dichotomous choice valuation question format has received considerable support (see Table 4.4). Valuation questions using dichotomous choice appear to elicit more consistent responses than open-ended questions or bidding games. Bishop and Heberlein's Wisconsin Sandhill study (1990) suggests that there is no significant difference between valuations collected from a hypothetical market using binary choice vs. actual cash transactions. At the same time, dichotomous choice questions lead to consistently higher estimates than open ended questions. Further research is necessary to determine the source of this error. The literature indicates a controversy surrounding CVM survey respondents' estimates of health risks and their comparability with expert opinion (Boyle et al., 1995; Boyle, 1994; Lazo et al., 1992). This controversy arises not only in the design of survey instruments but also in the use of results. It is generally acknowledged that individuals need a full information set that includes both general and specific information to identify their own best interests with respect to ground water protection programs (Poe and Bishop, 1993). Overly general information in the survey instrument appears to lead to biased estimates of willingness to pay. An illustrative example of the problems arising from differences between expert opinion and CVM estimates is sketched in Portney's 1992 study, where experts believed a chemical in ground water to be harmless, whereas citizens held the chemical responsible for above-average incidence of cancer and were willing to pay $1,000 for what experts say will be a costly and unnecessary treatment.

Very little empirical research has been devoted to establishing a minimum standard of information adequacy for CVM studies (Poe and Bishop, 1993; Powell, 1991; Boyle, 1994). The question remains as to what type of information should be presented to respondents and how that information affects estimated

TABLE 4.4 Summary of CVM Studies—Survey Characteristics

Author(s) Dates	Response Rates Usable Responses (Percent)	Payment Vehicle	Valuation Question
McClelland et al. (1992)	60 44	water bill	payment care
Caudill (1992) Caudill and Hoehn (1992)	67 60	higher taxes	dichotomous choice
Doyle (1991)	NA	bond	payment card
Edwards (1988)	78 58	bond	dichotomous choice, open ended
Jordan and Elnagheeb (1993)	35 34	water bill water purification equipment	payment card
Poe (1993); Poe and Bishop (1992)	76-91	increased taxes, lower profits, higher prices	dichotomous choice
Powell (1991) Powell and Allee (undated)	50	water bill higher taxes	payment card
Clemons, Collins, and Green (1995)	64	bond	dichotomous choice

SOURCE: Reprinted with permission from Boyle (1994).

values of the benefits of protecting ground water (Boyle, 1994). Lazo et al. (1992) provide guidelines for reducing information biases using verbal protocols.

The increasing costs of conducting benefit studies and decreasing support for research efforts have led to renewed efforts to minimize costs by establishing some method for transferring benefits from study sites to policy sites. Preliminary evidence (VandenBerg et al., 1995) indicates the challenges inherent in benefits transferability with ground water resources.

While complete transferability of benefits estimates is an impossible goal given the site-specific nature of most ground water valuations, the debate itself is leading to collaborative interdisciplinary efforts which may create benefits in and of themselves.

A final area, though not specifically an area of disagreement, is the dearth of

TABLE 4.5 A General Matrix of Ground Water Functions/ Services and Applicable Valuation Methods

Ground Water Function/Service Flow	Applicable Valuation Method
A. Extractive values	Cost of illness
1. Municipal use (drinking water)	Averting behavior
a) Human health - morbidity	Contingent valuation Contingent ranking/behavior
b) Human health - mortality	Averting behavior Contingent valuation Contingent ranking/behavior
2. Agricultural water use	Derived demand/production cost
3. Industrial water use	Derived demand/production cost
B. *In situ* values	
1. Ecological values	Production cost techniques Contingent valuation Contingent ranking/behavior
2. Buffer value	Dynamic optimization Contingent valuation Contingent ranking/behavior
3. Subsidence avoidance	Production cost Hedonic pricing model Contingent valuation Contingent ranking/behavior
4. Recreation	Travel cost method Contingent valuation Contingent ranking/behavior
5. Existence value	Contingent valuation Contingent ranking/behavior
6. Bequest value	Contingent valuation Contingent ranking/behavior

SOURCE: Adapted from Freeman, 1993. (Reprinted with permission from Resources for the Future, 1993. Copyright by Resources for the Future.)

information on nonuse values of ground water. For example, only one study (McClelland et al., 1992) has attempted to address existence value of ground water, and the approaches used in the study have been criticized. The estimates for existence and bequest values found in this study were smaller than use values found from other studies using indirect or direct methods. Additional research is needed to further document the existence and size of nonuse values for ground water resources. Table 4.5 illustrates the applicable valuation methods for addressing various potential ground water values. It is not an exhaustive list. Following Table 1.3, it is organized according to extractive and *in situ* services of ground water. Nonuse values are treated as a subcategory of *in situ* values in this scheme. Such values can only be measured using direct methods, such as CVM or a variant.

Two cautions should be kept in mind when examining Table 4.5. In some cases, several different methods can be used to measure the same ground water function. This permits the potential for checking the consistency of estimates of the same function or service. However, it also raises the potential that some decision-makers will double-count value estimates of the same service when attempting to arrive at a total value estimate of a particular ground water resource. Use of a comprehensive list of ground water functions and services can serve as a guide to keep correct calculations of total values from individual studies. Finally, the reader should recall the advantages and disadvantages of each of the techniques (summarized in Table 4.1) when considering their use in decision-making.

CONCLUSIONS AND RECOMMENDATIONS

- **For valid and reliable results to be obtained, the valuation method must be matched to the context and the ground water function or service of interest.**
- **It is hard to make generalizations about the validity and reliability of specific valuation approaches in the abstract. The validity of the approach depends on the valuation context and the type of ground water services in question. Different approaches are needed to value different services; care must be taken not to double count values resulting from different services.**
- **Previous ground water valuation studies have focused primarily on a small part of the known ground water functions and services (identified in Chapter 3). Thus current empirical knowledge of the values of ground water is quite limited and concentrated in a few areas, such as extractive values related to drinking water use.**
- **If data are available and critical assumptions are met, indirect valuation methods (e.g., TCM, HPM averting behavior) can produce reliable estimates of the use values of ground water.**
- **The contingent valuation method (CVM), when used correctly, has the potential for producing reliable estimates of ground water use values in certain contexts. However, few, if any, studies to date meet the stringent conditions, as established by a NOAA panel of Nobel-Laureate economists, that are required to produce defensible estimates of nonuse values. More research is needed to compare use values from CVM with those of other methods to determine whether CVM will consistently yield reliable estimates. CVM does have the advantage of allowing researchers to be precise in focusing on the total resource attribute to be valued, compared to the**

results from other indirect approaches that generally fail to capture total economic value.

• The EPA, and other federal agencies as appropriate, should develop and test valuation methods for addressing the use and nonuse values of ground water, especially considering the ecological services provided by ground water.

• Given the problems in using CVM to measure ground water values, EPA and other appropriate government agencies should encourage ways of enhancing the utility of CVM. For example, contingent ranking or behavior methods may be useful in improving the robustness of CVM estimates and may expand the potential for benefits transfer.

• Technical, economic, and institutional uncertainties should be considered and their potential influence delineated in ground water valuation studies. Research is needed to articulate such uncertainties and their potential influence on valuation study results.

• Ground water values obtained from both indirect and direct methods are dependent on the specific ground water management context. Attempts to generalize about or transfer values from one context to another should be pursued with caution.

• Traditional valuation methods such as cost of illness, demand/analysis, and production cost can be used for many ground water management decisions that involve use values. Such methods offer defensible estimates of what are likely to be the major benefits of ground water services.

• The pervasiveness and magnitude of nonuse values is uncertain. Few and limited studies have been conducted, and little reliable evidence exists with which to draw conclusions about the importance of nonuse values for ground water. Additional research is needed to document the occurrence and size of nonuse values for ground water systems.

• What is most relevant for decision-making regarding ground water policies or management is knowledge of how the TEV of ground water will be affected by a decision. Pending documentation of large and pervasive nonuse values for ground water, it is likely that in many, but not all, circumstances, measurement of use values or extractive values alone will provide a substantial portion of the change in TEV relevant for decision-making.

• In some circumstances the TEV is likely to be largely composed of nonuse values. At the current time, pending documentation of large and pervasive nonuse values for ground water systems, this appears to be most likely when ground water has a strong connection to surface water and a decision will substantially alter these service flows. In these situations, focusing on use values alone could seriously mismeasure changes in TEV and will ill serve decision-making. Decision-makers should approach valuation with a careful regard for measurement of TEV using direct techniques that can incorporate nonuse values.

REFERENCES

Abdalla, C. W. 1990. Measuring economic losses from ground water contamination: An investigation of household avoidance costs. Water Resources Bulletin 26(3):451-463.

Abdalla, C. W. 1994. Groundwater values from avoidance cost studies: Implications for policy and future research. American Journal of Agricultural Economics 76(5):1062-1067.

Abdalla, C. W., B. A. Roach, and D. J. Epp. 1992. Valuing environmental quality changes using averting expenditures: An application to ground water contamination. Land Economics 68(2):163-169.

Adamowicz, W. L. 1991. Valuation of environmental amenities. Canadian Journal of Agricultural Economics 39:609-618.

Arrow, K., R. Solow, E. Leamer, P. Portney, R. Randner, and H. Schuman. 1993. Report of the NOAA panel on contingent valuation. Federal Register 58(10):4602-4614.

Bain, J. S., R. E. Caves, and J. Margolis. 1966. Northern California's Water Industry. Baltimore: Johns Hopkins University Press.

Bartik, T. J. 1988. Evaluating the benefits of non-marginal reductions in pollution using information on defensive expenditures. Journal of Environmental Economics and Management 15:111-127.

Bergstrom, J. C., and J. H. Dorfman. 1994. Commodity Definition and Willingness-to-pay for Ground Water Quality Protection. Review of Agricultural Economics 16:413-425.

Berry, D. W., and G. W. Bonen. 1974. Predicting the municipal demand for water. Water Resources Research 10:1239-1242.

Bishop, R. C., and T. A. Heberlein. 1990. The contingent valuation method. In Economic Valuation of Natural Resources: Issues, Theory, and Applications. Rebecca L. Johnson and Gary V. Johnson, eds. Boulder: Westview Press.

Bishop, R. C., and M. P. Welsh. 1992. Existence values in benefit-cost analysis and damage assessment. Land Economics 68(4):405-417.

Boyle, K. J. 1994. A Comparison of Contingent-Valuation Studies of Ground Water Protection. Department of Resource Economics and Policy, Staff Paper 456, Maine Agricultural Experiment Station. Orono: University of Maine.

Boyle, K. J., and J. C. Bergstrom. 1994. A Framework for Measuring the Economic Benefits of Ground Water, Rep. 459. Maine Agricultural Experiment Station. Orono: University of Maine.

Boyle, K. J., M. P. Welsh, R. C. Bishop, and R. M. Baumgartner. 1995. Validating contingent valuation with surveys of experts. Agricultural and Resource Economics Review. 24(2):247-254.

Braden, J. B., and C. D. Kolstad. 1991. Measuring the Demand for Environmental Quality. Amsterdam: Elsevier Science.

Brown, T. C. 1984. The concept of value in resource allocation. Land Economics 60(3):231-246.

Burt, O. R. 1964. Optimal resource use over time with an application to groundwater. Management Science 11:80-93.

Burt, O. R. 1966. Economic Control of Groundwater Reserves. Journal of Farm Economics 48:632-647.

Carson, R. T. 1991. Constructed Markets in Measuring the Demand for Environmental Quality. J. B. Braden and C. S. Kolstad, eds. North Holland.

Caudill, J. D. 1992. The Valuation of Ground Water Pollution Policies: The Differential Impacts of Prevention and Remediation, Ph.D. dissertation. Department of Agricultural Economics, Michigan State University.

Caudill, J. D., and J. P. Hoehn. 1992. The Economic Valuation of Groundwater Pollution Policies: The Role of Subjective Risk Perceptions. Staff Paper No. 92-11. Department of Agricultural Economics, Michigan: Michigan State University.

Ciriacy-Wantrup, S. V. 1956. Concepts used as economic criteria for a system of water rights. Land Economics 32:295-312.

Clemons, R., A. R. Collins, and K. Green. 1995. Contingent Valuation of Protecting Groundwater Quality by a Wellhead Protection Program. Paper presented at the American Agricultural Economics Association Annual Meeting, Indianapolis, Indiana.

Cochrane, R., and A. W. Cotton. 1985. Municipal water demand study: Oklahoma City and Tulsa, Oklahoma. Water Resources Research 21:941-943.

Collins, A. R., and S. Steinback. 1993. Rural household response to water contamination in West Virginia. Water Resources Bulletin 29(2):199-209.

Courant, P. N., and R. C. Porter. 1981. Averting expenditures and the costs of pollution. Journal of Environmental Economics and Management 8(4):321-329.

Crutchfield, S. R., P.M. Feather, and D. R. Hellerstein. 1995. The benefits of protecting rural water quality: An empirical analysis. Agricultural Economic Report No. 701. Washington, D.C.: Economic Research Service, U.S. Department of Agriculture.

Diamond, P. A., and J. A. Hausman. 1994. Contingent valuation: Is some number better than no number? Journal of Environmental Perspectives 8(4):45-64.

Dickie, M., and S. Gerking. 1988. Benefits of Reduced Soiling from Air Pollution Control: A Survey. In Migration and Labor Market Efficiency. H. Folmer, ed. Amsterdam: Martin Nijhoff.

Doyle, J. K. 1991. Valuing the Benefits of Ground Water Cleanup. Interim report, Office of Policy, Planning and Evaluating U.S. EPA, Cooperative Agreement No. CR-815183.

Edwards, S. F. 1988. Option prices for groundwater protection. Journal of Environmental Economics and Management 15:475-487.

Fleming, R. A., R. M. Adams, and C. S. Kim. 1995. Regulating groundwater pollution: Effects of geophysical response assumptions on economic efficiency. Water Resources Research 31:1069-1076.

Foster, H. S., and B. R. Beattie. 1979. Urban residential demand for water in the U.S. Land Economics 55:43-58.

Freeman, A. M., III. 1993. The Measurement of Environmental and Resource Values: Theory and Methods. Washington, D.C.: Resources for the Future Press.

Hanemann, W. M. 1994. Valuing the environment through contingent valuation. Journal of Economic Perspectives 8(4):19-43.

Harrington, W., and P. R. Portney. 1987. Valuing the benefits of health and safety regulation. Journal of Urban Economics 22:101-112.

Hausman, J. 1993. Contingent Valuation: A Critical Assessment. Amsterdam: Elsevier Science.

Jordan, J. L., and A. H. Elnagheeb. 1993. Willingness to pay for improvements in drinking water quality. Water Resources Research 29:237-245.

Kelso, M. M., W. E. Martin, and L. E. Mack. 1973. Water Supplies and Economic Growth in an Arid Environment: An Arizona Case Study. Tucson: University of Arizona Press.

Kim, C. S., J. Hostetler, and G. Amacher. 1993. The regulation of groundwater quality with delayed responses: A multistage approach. Water Resources Research 29:1369-1377.

Kopp, R. J. 1992. Why existence values should be used in cost-benefit analysis. Journal of Policy Analysis and Management 11(1):123-130.

Lazo, J. K., W. D. Schulze, G. H. McClelland, and J. K. Doyle. 1992. Can contingent valuation measure non-use values? American Journal of Agricultural Economics (December):1126-1132.

Malone, P., and R. Barrows. 1990. Groundwater pollution's effects on residential property values, Portage County, Wisconsin. Journal of Soil and Water Conservation 45(2):346-348.

Martin, R. C., and R. P. Wilder. 1992. Residential demand for water and the pricing of municipal water services. Public Finance Quarterly 20:93-102.

McClelland, G. H., W. D. Schulze, J. K. Lazo, D. M. Waldman, J. K. Doyle, S. R. Elliott, and J. R. Irwin. 1992. Methods for Measuring Non-Use Values: A Contingent Valuation Study of Groundwater Cleanup. Boulder: University of Colorado.

McFadden, D. 1994. Contingent valuation and social choice. American Journal of Agricultural Economics 76:689-708.

Mendelsohn, R., and D. C. Markstrom. 1988. The Use of Travel Cost and Hedonic Methods in Assessing Environmental Benefits. In Amenity Resource Valuation: Integrating Economics with Other Disciplines. G. L. Peterson, B. L. Driver, and R. Gregory, eds. State College, Penn.: Venture Publishing.

Mitchell, R. C., and R. T. Carson. 1989. Using Surveys to Value Public Goods: The Contingent Valuation Method. Washington, D.C.: Resources for the Future Press.

Musser, W. N., L. M. Musser, A. S. Laughland, and J. S. Shortle. 1992. Contingent Valuation and Averting Costs Estimates of Benefits from Public Water Decisions in a Small Community. Agricultural Economics and Rural Sociology Staff Paper No. 171. University Park: Pennsylvania State University.

National Oceanic and Atmospheric Administration. 1995. Natural resource damage assessments. Federal Register 60(149):39804-39834.

Page, G. W., and H. Rabinowitz. 1993. Groundwater contamination: Its effect on property values and cities. Journal of the American Planning Association 59(4):473-481.

Peterson, G. L., C. Sorg, D. McCollum, and M. H. Thomas, eds. 1992. Valuing Wildlife Resources in Alaska. Boulder: Westview Press.

Poe, G. L. 1993. Information, Risk Perceptions and Contingent Values: The Case of Nitrates in Groundwater. Ph.D. dissertation. Department of Agricultural Economics, University of Wisconsin.

Poe, G. L., and R. C. Bishop. 1993. Prior information, general information, and specific information in the contingent valuation of environmental risks: The case of nitrates in groundwater. Cornell Agriculture Economics Staff Paper 93-11.

Poe, G. L., and R. C. Bishop. 1992. Measuring the benefits of groundwater protection from agricultural contamination: Results from a two-stage contingent valuation study. Staff Paper No. 341. Madison: Department of Agricultural Economics, University of Wisconsin-Madison.

Portney, P. R. 1994. The contingent valuation debate: Why economists should care. Journal of Economic Perspectives 8(4):3-17.

Portney, P. R. 1992. Trouble in Happyville. Journal of Policy Analysis and Management 11(1):131-132.

Powell, J. R. 1991. The Value of Ground Water Protection: Measurement of Willingness-to-Pay Information and Its Utilization by Local Government Decisionmakers. Ph.D. dissertation. Cornell University.

Powell, J. R., and D. J. Allee. Undated. The Estimation of Groundwater Protection Benefits and Their Utilization by Local Government Decision Makers. Final Report of a project conducted September 1988-September 1990. United States Geological Survey, Department of the Interior. Award Number 14-08-0001-G1649.

Randall, A. 1991. Total and Non-use Values in Measuring the Demand for Environmental Quality. J. B. Braden and C. D. Kolstad, eds. Amsterdam: Elsevier Science.

Renzetti, S. 1992. Evaluating the welfare effects of reforming municipal water prices. Journal EEM 22:147-163.

Rosenthal, D., and R. H. Nelson. 1992. Why existence values should not be used in cost-benefit analysis. Journal of Policy Analysis and Management 11(1):116-122.

Shultz, S. D. 1989. Willingness to Pay for Ground Water Protection in Dover, NH: A Contingent Valuation Approach. M.S. thesis. University of New Hampshire.

Shultz, S. D., and B. E. Lindsay. 1990. Willingness to pay for ground water protection. Water Resources Research 26(9):1869-1875.

Shultz, S. D., and A. E. Luloff. 1990. The threat of nonresponse bias to survey research. Journal of the Community Development Society 21(2):104-115.

Smith, V. K. 1989. Taking stock of progress with travel cost recreation demand methods: Theory and implementation. Marine Resource Economics 6:279-310.

Smith, V. K. 1993. Nonmarket valuation of environmental resources: An interpretive appraisal. Land Economics 69:1-26.

Smith, V. K., and W. H. Desvouges. 1986. Averting behavior: Does it exist? Economics Letters 20:291-296.

Snyder, J. H. 1954. On the economics of groundwater mining. Journal of Farm Economics 36:600-610.

Spofford, W. O., A. J. Krupnick, and E. F. Wood. 1989. Uncertainties in estimates of the costs and benefits of groundwater remediation: Results of a cost-benefit analysis. Discussion Paper QE 89-15. Washington, D.C.: Resources for the Future Press.

Stevens, T. H. 1991. Measuring the existence value of wildlife: What do CVM estimates really show? Land Economics 67:390-400.

Sun, H. 1990. An Economic Analysis of Ground Water Pollution by Agricultural Chemicals. M.S. thesis. University of Georgia.

Sun, H., J. C. Bergstrom, and J. H. Dorfman. 1992. Estimating the benefits of groundwater contamination control. Southern Journal of Agricultural Economics 24(2):63-71.

Teeples, R. K., and D. Glyer. 1987. Production functions for water delivery systems: Analysis and estimation using dual cost function and implicit price specification. Water Resources Research 23:765-773.

Tsur, Y., and T. Graham-Tomasi. 1991. The buffer value of groundwater with stochastic surface water supplies. Journal of Environmental Economics and Management 21:201-224.

VandenBerg, T. P., G. L. Poe, and J. R. Powell. 1995. Assessing the accuracy of benefits transfers: Evidence from a multi-site contingent valuation study of groundwater quality. Working Paper 95-01, Department of Agricultural, Resource, and Managerial Economics, Cornell University.

Wong, S. T. 1972. A model of municipal water demand: a case study on northeastern Illinois. Land Economics (48):34-44.

5

Legal Considerations, Valuation, and Ground Water Policy

This chapter outlines the fixed ground water allocation and quality policies that affect ground water valuation. The chapter also describes how new policies are trying to balance environmental protection with the corresponding economic consequences. It concludes with a section that addresses research needs based on the lack of information relating valuation information to ground water management decision. American ground water policy is a combination of state, local and federal laws dealing with ground water allocation and ground water quality protection. Ground water allocation is almost exclusively the province of state law, whereas ground water quality protection is a mixture of state, local, and federal laws. State ground water allocation laws date back to 19th-century court decisions (Murphy, 1991; Tarlock, 1995), while most modern ground water quality laws date from the 1970s and 1980s (Beck, 1991).

An important ground water allocation issue is how to evaluate current versus future use of ground water. Unfortunately, states rarely consider future ground water uses in establishing ground water allocation policies dealing with ground water depletion. The states that do have explicit policies to limit ground water depletion typically simply prohibit additional ground water uses and do little to regulate current ground water uses to extend aquifer life (Aiken, 1982). There is unfortunately too little attention given to regulating existing ground water uses to lengthen aquifer life, let alone any explicit quantitative evaluation of the trade-off between current and future ground water use. Consequently, ground water valuation has historically played almost no role in state ground water allocation policies. Ground water policies in most states could be strengthened by acknowledging ground water's future value.

Valuation has played a more significant role in ground water quality policies, however. Under state and federal Superfund programs, valuation is a critical factor in determining the amount of money recoverable for natural resources damages. Valuation has also implicitly (through cost-effectiveness analyses) become a significant factor in determining ground water remediation levels in Superfund cleanup sites and future drinking water standards.

Valuation may become an even more important consideration in ground water protection policy if formalized regulatory benefit-cost studies become a more significant part of developing and selecting federal environmental regulations. Critics of current federal environmental policies contend that those policies should have as their primary objective risk reduction rather than environmental protection. Focusing environmental regulations on risk reduction and subjecting proposed environmental regulations to regulatory benefit-cost studies has been required administratively since the Carter administration. Proposed federal legislation would elevate risk reduction and regulatory cost-benefit tests to the highest environmental policy objectives, raising the policy significance of resource valuation. Such a change would represent a fundamental shift in federal environmental policy.

VALUATION AND GROUND WATER ALLOCATION

As stated earlier in this report, most states treat ground water as if it were a free good: well owners are subject to few if any pumping restrictions except in unusual circumstances. Where no use restrictions are established, ground water tends to be undervalued. In most eastern states ground water use is subject only to court decisions, which effectively means there are no legal constraints on ground water development and use. Eastern and western states that require state permits as a condition of ground water use seldom impose pumping limits that take aquifer life into account. Only rarely are current ground water uses balanced against future water needs.

Ground Water Allocation Law Doctrines

Ground water law in the United States is the result of a bewildering mix of state court decisions and state statutes. While some generalization is possible, each state's ground water law is unique. This overview broadly categorizes state ground water laws primarily as they relate to allocation, and briefly surveys state and federal ground water protection policies.

Common Law States

The common law doctrines of absolute ownership, reasonable use, correlative rights, and eastern correlative rights are based on state court decisions and are implemented through litigation or private negotiation. While prior appropriation was initially adopted in a few western states by court decision, it will be discussed separately as a statutory rather than a judicial doctrine.

Absolute Ownership. The earliest judicial theory of ground water rights is the doctrine of absolute ownership, also referred to as the English rule. Under the absolute ownership doctrine the landowner is, by virtue of land ownership, considered owner of the ground water in place. Thus in absolute ownership jurisdictions a landowner may pump as much ground water as possible, without regard to the effect his pumping has on neighboring landowners.

The English rule of absolute ownership reflected 19th-century judicial observations that the movement of ground water was unknowable and thus it was unfair to hold a landowner liable for interfering with a neighbor's well when it was not knowable whether the defendant's pumping actually affected the plaintiff's well or not. The English rule was once quite popular in the United States, but now only Texas among the western states remains an absolute ownership jurisdiction; although Texas now has a number of sub-state districts where ground water use is now regulated, e.g. Houston/Galveston area, High Plains, San Antonio. Some eastern states may still be English rule jurisdictions, but the judicial trend is toward adoption of the eastern correlative rights doctrine.

Reasonable Use. The reasonable use rule, or American rule, was developed in the 19th century. Under the American rule landowners are entitled to use ground water on their own land without waste. If their use exceeds this "reasonable use," the landowner is liable in damages. The American rule may still be followed in a few eastern states, although it is being judicially replaced by the eastern correlative rights doctrine. The reasonable use doctrine is part of the ground water jurisprudence of Nebraska, Arizona, and California.

Western Correlative Rights. The California doctrine of correlative rights also initially developed in the 19th century but has continued to develop. Under the correlative rights doctrine, if the ground water supply is inadequate to meet the needs of all users, each user can be judicially required to proportionally reduce use until the overdraft is ended. The policy significance of correlative rights is that each well owner is treated as having an equal right to ground water regardless of when first use was initiated.

The correlative rights doctrine is part of the ground water jurisprudence of California and Nebraska, although its sharing feature has been incorporated into the ground water depletion statutes of a few other western states.

Eastern Correlative Rights. The eastern correlative rights doctrine, inspired by the Second Restatement of Torts, states that when conflicts between users occur, water will be allocated to the "most beneficial" use, giving consideration to a wide variety of factors, including priority of use. Several factors are enumerated to be considered in a judicial determination of whether a water use at issue is "unreasonable": (1) the purpose of the interfering use, (2) the suitability of the interfering use to the watercourse, (3) the economic value of the interfering use, (4) the social value of the interfering use, (5) the extent and amount of harm it causes, (6) the practicality of avoiding the harm by adjusting the use or method of use of one riparian proprietor or the other, (7) the practicality of adjusting the quantity of water used by each proprietor, (8) the protection of existing values of water uses, land, investments, and enterprises, and (9) the justice of requiring the user causing the harm to bear the loss.

Statutory States

Permit States. A few states, including Florida, Iowa, Wisconsin, and Minnesota, require a state permit as a condition of well construction and use. Typically, users become subject to a rationing program during periods of shortage so that public water supplies are protected at the expense of other uses.

Appropriation States. With the exception of the major ground water using states (e.g., Texas, Nebraska, Arizona, and California), western states apply the doctrine of prior appropriation to ground water. This means that the right itself is dependent upon obtaining a state permit rather than simply owning land overlying the ground water supply. Between ground water users, priority of appropriation gives the better right. This means that first in time is first in right.

Ground Water Allocation Law Issues

Ground water rights conflicts between well owners, and problems caused by surface-ground water interference are among the issues addressed by allocation law issues.

Ground Water Rights

In the common law states, ground water rights are based upon owning land overlying the ground water supply and are defined by court decision. In the statutory states, including the eastern permit states, ground water rights are based upon obtaining a state permit and complying with its terms. In the permit and appropriation states, state statutes generally define the extent of ground water rights.

Common Law States. In all common law states, the right to use ground water is based on owning land overlying the ground water supply. In absolute ownership states, pumping is not restricted to avoid harm to others or to avoid waste.

(However, in Texas, malicious pumping may be judicially restrained.) In reasonable use jurisdictions there is generally no ownership interest in the ground water itself until it has been captured. Pumping may be judicially restrained to prevent waste or non-overlying uses. In correlative rights jurisdictions the right to use ground water is also based on owning land overlying the ground water supply, although in California prescriptive rights can be obtained for nonoverlying uses. Pumping may be judicially restrained to prevent waste or to apportion an inadequate supply. In eastern correlative rights states, pumping may be judicially restrained during shortages, although the basis upon which shortages will be allocated is not predictable. Ground water rights are least well defined in the eastern correlative rights statutes, since judicial notions of what may constitute the "most beneficial" use of ground water may change over time.

Statutory States. In both eastern permit states and appropriation states, rights to use ground water are based on obtaining and complying with the terms of a state permit. However, most existing ground water uses were automatically grandfathered into the permit system. Pumping rates may be limited in a permit and further limited during shortages. In eastern permit states, public water supply uses and domestic uses will generally be protected during shortages. In appropriation states, senior users (i.e., those with an earlier priority date, or in other words, an older well) are protected during shortages without regard to use. A junior user with a higher use, however, may be able to condemn a senior's use right during shortages and thus pump water out of priority.

Well Interference Conflicts

Well interference is where the cone of depression of one well intersects with the cone of depression of another well, reducing the yield of both wells. In an artesian aquifer, well interference may occur when the pumping from one well drops the water level below the pump of another well. Well interference may occur even when there is sufficient water available to supply all users—it may be the result of inadequate wells rather than an inadequate supply. Most ground water disputes have tended to be well interference disputes.

Common Law States. In absolute ownership states, a landowner is not liable for interfering with a neighbor's well. Thus the neighbor's only recourse is to drill a new well deeper than the neighbor's well. This has been described as "the race to the pumphouse." In reasonable use states a landowner complaining of well interference is entitled to relief only if the complained-of use is wasteful or not on overlying land. Thus plaintiffs complaining of well interference have little legal remedy in the absence of gross waste or nonoverlying uses. The courts' definition of what constitutes a wasteful use is rather generous. Arizona courts have defined overlying land to include only the tract of land where the well is located. Nebraska, a reasonable use state, minimizes well interference conflicts between high-capacity wells through statutory well-spacing restrictions. In correlative rights states, competing pumpers have equal rights during shortages.

Statutory States. In eastern permit states and appropriation states, well interference conflicts may be reduced through permit conditions such as well-spacing restrictions and pumping restrictions. Further, in eastern permit states public water supply and domestic uses are generally protected during shortages. Otherwise, correlative rights principles will likely be applied.

Prior appropriation is primarily a surface water doctrine that has been applied rather uncritically to ground water. As ground water problems developed, the principles of prior appropriation were modified to better apply to the ground water context. Two modifications that were made in response to well interference conflicts are establishment of reasonable pumping depths and problem area regulations.

Sometimes the senior or oldest wells may not be fully penetrating. To allow senior appropriators to insist upon original pumping depths being maintained could seriously constrain ground water development. Thus several appropriation states do not strictly maintain priority during well interference disputes, but only protect "reasonable pumping depths" through well permit restrictions on pumping. If a senior's well cannot pump at that depth, typically the senior appropriator is responsible for replacing the well.

In some appropriation states ground water development and use has resulted in chronic well interference problems. Special pumping and development restrictions may be imposed by the state engineer in designated problem areas. Regulations can include a ban on new high-capacity wells and pumping restrictions to maintain reasonable pumping depths and reduce interference conflicts.

Ground Water Depletion

Ground water depletion occurs when withdrawals from the aquifer exceed recharge. This is sometimes referred to as ground water overdraft or mining. Overdraft is significant in the Ogallala aquifer region, including Nebraska, Kansas, Colorado, Texas, and New Mexico, as well as in California, Nevada, and Arizona.

The amount of water that can be safely withdrawn without leading to long-term aquifer depletion is sometimes referred to as the safe-yield amount.

Common Law Doctrines. Of the overlying rights doctrines, only the correlative rights doctrine addresses depletion. In absolute ownership states pumpers can completely ignore depletion and in reasonable use states they need be concerned about depletion only to the extent that their uses are wasteful or nonoverlying. In eastern correlative rights states, courts can apportion water between competing users. Florida (a permit state) is the primary eastern state with significant ground water depletion concerns.

In theory, courts in correlative rights states can limit withdrawals to an aquifer's safe yield, thus preventing depletion. In practice, in California safe-yield adjudications are used primarily to define baseline pumping rights so that ground water recharge agencies can charge pumpers a pumping fee for using more than their safe-yield allocation.

Problem Area Regulations. Dealing with ground water depletion has not been a major feature of eastern permit state ground water administration. Most ground water shortages are seasonal rather than perennial. In appropriation states, the most common way to deal with depletion is to establish special problem area regulations. Once the problem area has been administratively defined, typically no new high-capacity wells may be drilled within the problem area. Less frequently are the uses of existing appropriators limited, a significant policy failing. Initial ground water appropriation allocations are typically generous, not requiring a high degree of water use efficiency. Where problem area allocations have been established, they typically are high enough to allow current irrigation practices to be maintained with little or no change. Any increases in irrigation efficiency typically come only as well yields decline.

Nebraska and Arizona, both reasonable use states, have adopted problem area statutes to deal with ground water depletion. Through statute and regulation, Arizona identified areas with depletion problems and instituted water-use reductions in phases. Nebraska's depletion statute is a local option for natural resource districts (NRDs). In one area the NRD has required well metering and pumping reductions. Most NRDs, however, have opted to ignore their depletion problems. Texas, an absolute ownership state, has a similar local option approach in which runoff irrigation controls and education programs have been implemented by local ground water conservation districts.

Conjunctive Use. In California, the courts have recognized the rights of entities storing water underground to control the use of that water. As a result, when ground water pumpers have received their safe-yield allocation through a court adjudication, they are typically required to pay a fee to the recharge entity pumping water stored underground, i.e., for pumping ground water in excess of their safe yield allocation. Where both surface water and ground water are available to ground water pumpers, the recharge entity can raise or lower ground water pumping fees to encourage surface water use during periods of ample surface supplies or to discourage surface water use during periods of surface water shortage.

Surface-Ground Water Interference

Where ground and surface water supplies are hydrologically connected, courts typically have followed the "underground stream" doctrine to interrelate surface and ground water rights of use. This means that wells will be treated as surface diversions and governed by surface water law. Riparian rights jurisdictions basically follow a theory similar to the reasonable use doctrine of ground water rights, although the eastern correlative rights doctrine has been applied to both surface and ground water. Where surface water rights are appropriative, priority would govern surface-ground water disputes.

Under the "Templeton" doctrine, the New Mexico State Engineer has required a junior ground water appropriator to purchase and retire sufficient surface appropriations to compensate for the expected stream depletion effect of a proposed well. Colorado has an elaborate system for integrating surface appropriations and appropriations of subflow and tributary ground water. Generally junior ground water appropriators are expected, through plans of augmentation, to compensate the stream for their expected stream depletion effects of well pumping.

Ground water pumping that reduces stream flows upon which endangered species depend for habitat may be regulated under the federal Endangered Species Act. In Texas, the Sierra Club has sued the U.S. Fish and Wildlife Service, arguing that well pumping from the Edwards Aquifer near San Antonio has reduced stream flows to the detriment of endangered species. Municipal and irrigation uses of ground water are currently under federal court order to be reduced to meet endangered species stream flow requirements.

In some instances, prevailing ground water laws impose constraints on individuals and institutions from assigning the appropriate TEV to ground water in operationally meaningful ways. Among the constraints that can be found in prevailing ground water law are: failures to vest clearly the right to use ground water (i.e., provisions which allow the law of capture to prevail); prohibitions or constraints on the marketing of ground water; forfeiture provisions in which the failure to apply the water to beneficial use may risk forfeiture of right; provisions which encourage individual users to ignore the social costs of use; and provisions which encourage the exploitation of ground water in a competitive fashion (virtually all of these barriers to accurate depiction of TEV apply to surface water as well).

Water Rights Transfers

In most western states, where water supplies are short, irrigation is the predominant use. To accommodate new uses for municipal, industrial, recreational, and environmental purposes, water rights in most western states can be bought and used for a different purpose. Typically this involves quantifying the original consumptive use, and allowing that amount to be transferred to another party so it can be used for a different purpose and perhaps in a different location. Restricting transfers to consumptive use preserves return flows from the original use to other appropriators on the stream.

Water marketing has been hailed as the modern, environmentally friendly way to deal with water shortages in the arid West, as opposed to dam construction. Most water right transfers involve surface water rights. However, ground water transfers are common in Arizona. Water markets provide flexibility in water use and management while also providing "real world" prices for water, which can be useful in attempting to value ground water. More states should consider the authorization and promotion of water marketing, including transfer of ground water rights when appropriate.

An emerging policy issue is how to deal with adverse community impacts from transferring water from irrigation to nonagricultural uses (NRC, 1992). A principal concern is that as water is transferred away from irrigated agriculture to other uses, the community's agribusiness economic base may be threatened.

A variation on the water right transfer theme is for an entity, typically referred to as a water bank, to purchase water rights from users willing to sell them, and then resell the water to whoever needs it. Water right sales to a water bank may be temporary, whereas most water right transfers are permanent.

Water marketing and banking are important to valuation in that such water right transfers provide actual market values for water, which in turn provide crucial information for valuing ground water.

Valuation and Ground Water Management

Arizona's Ground Water Management Act of 1980 reflects a series of conscious water allocation choices to a much greater degree than most state water allocation legislation does. Arizona's Act mandates the goal of eliminating ground water overdraft by the year 2025. Overdraft is to be reduced by a series of 5- and 10-year plans that apply to Arizona's most populated areas and agricultural center. Thus Arizona's Ground Water Management Act is a rare example of state legislation that implicitly values ground water. Such legislation can help pave the way for other states to use valuation studies in determining ground water's future economic worth. Florida also regulates ground water uses to protect public water values, including environmental services from ground water.

The principal consequence of the law of capture ground water allocation policy relied on by many states is that potential future uses of ground water are not taken into account. Valuation plays no significant role in ground water allocation policy under the law of capture. Indeed, the extent to which state water law fails to deal effectively with ground water depletion indicates the degree to which its policies ignore valuation.

Valuation as an analytical tool has typically been more important to water suppliers in helping them evaluate water supply alternatives. A recent study involving ground water valuation was prepared for the city of Albuquerque to help it evaluate water supply alternatives, it is described further in Chapter 6 (Brown et al., 1996).

VALUATION AND GROUND WATER QUALITY PROTECTION

Ground water quality protection law is perhaps even more fragmented than ground water allocation law. While federal law provides a legal umbrella, no single unifying federal ground water quality protection law exists. The federal Clean Water Act does not establish a basic program for ground water quality protection the way it does for surface water. Rather, federal ground water quality protection programs are scattered throughout a variety of federal environmental statutes.

For example, resource valuation plays a significant role in Superfund policy. First, valuation is implicitly involved in benefit-cost analyses of Superfund cleanup alternatives. Second, valuation is critical in recovering damages caused to natural resources. Under the 1996 Safe Drinking Water Act amendments, benefit-cost analysis will be used in establishing new drinking water standards.

Ground Water Quality Protection

Ground water quality protection measures generally address either point source or non-point source pollution.

Point Source Pollution

CWA. Section 502(14) of the Clean Water Act (CWA) defined a point source of water pollution to include "any discernible, confined and discrete conveyance. . . from which pollutants are or may be discharged." This would include a discharge pipe into a stream or an injection well. The CWA thoroughly regulates point discharges into waters of the United States, including streams and wetlands. However, the CWA does not regulate point source discharges to ground water.

SDWA. The Safe Drinking Water Act requires public water suppliers to periodically test the quality of the drinking water they supply to their customers. If testing reveals violations of one or more of the EPA drinking water standards (also referred to as maximum contaminant levels or MCLs), remedial action must be taken.

Under the 1996 SDWA amendments, new MCLs will be established by EPA at a slower pace, and they will be subject to benefit-cost analyses and risk assessment. (More on this later.) States can grant waivers for drinking water system violations to communities of up to 3,300 people, and to communities up to 10,000 with EPA approval.

The SDWA has three provisions relating to ground water. First, MCLs are used by EPA as reference points in determining Superfund ground water cleanup standards. Second, EPA may designate aquifers as sole source aquifers for drinking water supply and prohibit federal activities adversely affecting the sole source aquifer. Finally, underground waste injection is regulated through the SDWA underground injection control program.

A 1996 amendment to the SDWA may further affect ground water quality protection. The 1996 amendments provide funding for communities to enter into source protection programs to protect the community's drinking water quality from contamination. Where ground water is the source of a community's water supply, SDWA funding will be made available for ground water quality protection.

CERCLA. The federal Superfund program of the Comprehensive Environmental Response, Compensation, and Liability Act (CERCLA), and the Superfund Amendments and Reauthorization Act (SARA), requires cleanup of ground water contaminated by waste disposal. Ground water remediation may be required to comply with MCL standards, although less stringent cleanup standards may be approved by EPA on a case-by-case basis through a technical waiver process.

The issue of "how clean is clean" under CERCLA is very contentious. In addition, significant CERCLA effort is expended to receive or recoup cleanup costs from responsible parties and potentially responsible parties. Many have argued that a more no-fault approach would result in faster cleanup than the current approach, which encourages litigation.

RCRA. The Resource Conservation and Recovery Act (RCRA) provides for cradle to grave regulation of hazardous waste transport and disposal. In addition, underground storage tanks (USTs) storing petroleum products and other hazardous chemicals, are regulated under RCRA. Cleanups from leaky underground tanks may be paid for through state UST cleanup funds if the leaky tank complies with cleanup eligibility requirements (which vary considerably among states).

Private Cleanup Liability. In addition to ground water cleanup liability imposed under CERCLA and RCRA, those guilty of ground water contamination may be liable under state cleanup statutes or court decisions for the costs of ground water cleanup. Most states have state Superfund laws for ground water cleanup as well as state UST cleanup funds. However, state laws vary considerably regarding the degree of financial responsibility imposed, ranging from a no-fault approach to a CERCLA approach.

Nonpoint Source Pollution

The term nonpoint source water pollution is not defined in the CWA. A draft EPA guidance document, however, defines NPS pollution:

> NPS pollution is caused by diffuse sources that are not regulated as point sources and normally is associated with agricultural, silvicultural and urban runoff, runoff from construction activities, etc. Such pollution results in human-made or human-induced alteration of the chemical, physical, biological, and radiological integrity of water. In practical terms, nonpoint source pollution does not result from a discharge at a specific, single location (such as a single pipe) but generally results from land runoff, precipitation, atmospheric deposition, or percolation. Pollution from nonpoint sources occurs when the rate at which pollutant materials entering water bodies or ground water exceeds natural levels.

The CWA does not directly regulate NPS (which can be a significant source of ground water contamination). However, EPA encourages states to control NPS pollution of surface and ground water through the Section 319 program (successor to the Section 208 program). EPA also requires states to implement NPS control strategies through the Coastal Zone Management Act (CZMA). Many observers expected EPA NPS authorities to be expanded in the anticipated CWA reauthorization along the lines of the CZMA NPS authorities. The CWA will not be reauthorized before 1997 at the earliest, however, and whether EPA NPS authorities will be extended beyond the current Section 319 program is uncertain.

FIFRA. EPA has broad authority to regulate and prohibit pesticide use under the Federal Insecticide, Fungicide and Rodenticide Act (FIFRA). Most states are also authorized by state statutes to regulate pesticide applications to protect ground water quality. In addition, EPA, under its FIFRA Pesticides in Ground Water Strategy, is requiring states to prepare management plans to further regulate pesticide use to protect ground water quality when EPA determines that ground water quality will not be adequately protected by simply following pesticide label directions.

Fewer states are authorized to regulate fertilizer applications to protect ground water quality. EPA currently lacks legal authority to regulate fertilizer application to protect ground water quality.

Prevention of Ground Water Contamination

EPA's principal strategy for dealing with ground water pollution is to prevent its occurrence, due principally to the very high costs of ground water remediation. Through grants to states, EPA can encourage ground water pollution prevention despite EPA's lack of a general ground water protection authority.

Watershed Management. Under Section 319 of the CWA, EPA makes grants to states to address surface and ground water NPS pollution on a watershed basis. States typically make cost-sharing funds available to farmers to implement conservation measures to reduce erosion and sedimentation, and to implement agricultural chemical best management practices (BMPs).

Wellhead Protection. Under the SDWA, EPA provides some funding to states to support local wellhead protection (WHP) programs. States help public water suppliers identify the WHP area, typically the 20-year time of travel area for the community's wellfield. Once the WHP area has been identified, communities are encouraged to inventory potential contaminant sources and to use their zoning authorities to keep incompatible land uses and practices outside of the WHP area. Existing incompatible uses are encouraged to take extra precautions to prevent contamination or to relocate outside the WHP area.

Planning and Zoning. Local communities and counties may exercise their zoning authorities to regulate or limit land uses that may contaminate ground water. Some states may use their facility licensing authority to protect ground water when licensing the location of hazardous waste disposal and similar facilities.

Valuation and Superfund Site Cleanup

The original 1980 Superfund ground water remediation policies were cost insensitive: cleanup cost was not heavily considered in establishing site cleanup standards. This changed in the 1986 Superfund amendments (SARA). Under the current federal Superfund program, remediation actions must be "cost-effective over the period of potential exposure or contamination" (42 U.S.C.A. 9605(a)(7); Seiver, 1996). Thus as a basic principle, ground water contamination cleanups must be "cost effective." Of course, cleanup standards also significantly influence cleanup costs. In addition, EPA must establish national cleanup priorities for potential cleanup sites. Cleanup priority criteria include: (1) the affected population, (2) the specific health risk associated with the hazardous materials to be remediated, (3) the potential for direct human contact, (4) destruction of sensitive ecosystems, and (5) natural resource damage affecting the human food chain (42 U.S.C.A. (a)(8)(A)). These cleanup priority criteria suggest that some ground water may be more valuable than other ground water based on the economic and environmental demand for the ground water.

Perhaps the most controversial Superfund issue revolves around quality objectives. Ground water, for example, is generally required to be cleaned up to drinking water standards, regardless of the expected future use of the water. Critics contend that such policies ignore the real, more limited risk of the contaminated costs, and drive up cleanup costs to the point that the Superfund program may face bankruptcy (NRC, 1994). EPA may, however, relax ground water cleanup standards for particular cleanup sites from 10^{-6} (i.e., the estimated health risk is one additional cancer out of 1 million persons exposed over 70 years) to 10^{-4} (i.e., the estimated health risk is one additional cancer out of 10,000 persons exposed over 70 years). Accepting the higher health risk would significantly lower cleanup costs. The fundamental policy issue is the following: What level of contamination represents an acceptable risk?

Valuation and Natural Resource Damages

Superfund and other federal environmental programs (Olson, 1989) authorize states and the federal government, among others, to sue (as "natural resources trustees") polluters for damages for natural resources destroyed or injured by hazardous substance releases and oil spills. Superfund also requires federal regulations to be developed for assessing the value of injured or destroyed natural resources. Any natural resource damage assessments conducted by trustees and following the regulations adopted by the U.S. Department of the Interior receive a rebuttable presumption of accuracy. Trustees may use other valuation techniques, but the damage assessment would not enjoy a rebuttable presumption of correctness.

The Department of the Interior adopted its natural resource damage assessment regulations in 1986 and 1987 (Copple, 1995). The damage assessment regulations were invalidated in part in federal court in 1989. Among other things, the court ruled that damages recovered by trustees should not be limited to the lesser of (1) the cost of restoring or replacing the equivalent of the injured resource or (2) the lost use value of the resource. However, the court did sustain the use of the contingent valuation method in natural resource damage assessment. Amendments to Department of the Interior regulations on damage assessment regulations are pending.

CHANGING ENVIRONMENTAL PRIORITIES: POLICY DIMENSIONS OF GROUND WATER VALUATION

A fundamental environmental policy issue is whether pollution control or other environmental regulations are health-based or technology-based. Health-based regulations begin with the premise that once a desired level of environmental quality has been determined, polluting activities will be regulated to whatever extent necessary to accomplish the specified environmental quality. Thus the resulting pollution control requirements are established (1) without regard to the availability of pollution control technology to accomplish the specified environmental standard and (2) without regard to the economic costs imposed on polluting entities. The health-based approach is favored by those placing a high priority on environmental protection. Critics, however, contend that health-based environmental standards disregard the economic cost of environmental controls and are likely to result in pollution controls too costly relative to the environmental benefits they yield.

Technology-based regulation offers a different model. Under this approach polluting entities are required to adopt currently available pollution control technology that is affordable. Whether adoption of this technology yields the desired level of environmental quality remains secondary to technological and economic feasibility. Technology-based regulations are favored by those subject to pollution control requirements. Critics of technology-based regulations contend they will not reduce pollution levels sufficiently to protect human health and the environment. Economists have also criticized the technology-based approach as being economically inefficient: the same amount must be spent on pollution control regardless of the actual amount of pollution being abated.

Critics of current environmental policy contend that even technology-based regulations pay too little attention to the economic costs imposed by environmental regulations. Beginning with the Reagan administration, increasing efforts have been made to explicitly quantify the economic and environmental costs and benefits of proposed environmental regulations to balance those costs and benefits. Regulatory impact assessments (RIAs) of proposed EPA environmental control regulations have been conducted to determine whether the economic and environmental benefits of proposed regulations justify the regulations' proposed costs. Similar analyses are being conducted to see whether environmental regulators are focusing their attention and resources on those problems posing the greatest risk to human health and the environment.

The RIA and similar evaluation processes require a more explicit balancing of often competing environmental and economic objectives in the regulatory process. Critics of current environmental policies are forcefully contending that those policies should focus on reducing risk and should impose only those regulations that can achieve their objectives in a cost-effective way.

Regulatory Impact Assessment

One controversial environmental policy issue is the extent to which federal pollution control requirements should balance protecting human health and the environment with compliance costs. Most federal environmental laws are cost insensitive, subordinating compliance costs to the protection of human health and the environment. Only a few federal laws require formal balancing of environmental protection and compliance costs. Critics contend that the costs of many pollution control (and similar) programs impose costs that are not commensurate with the environmental benefits achieved; that billions of dollars are spent to guard against relatively trivial risks.

President Reagan initiated a formal balancing of environmental protection and regulatory compliance costs through his controversial Executive Order 12291, which required EPA and other agencies to prepare regulatory benefit-cost analyses for proposed regulations imposing public and private costs of at least $100 million annually. Agency benefit-cost analyses were reviewed by the Office of Management and Budget (OMB). Environmentalists opposed Executive Order 12291 as administratively imposing benefit-cost criteria on federal environmental rules when such criteria were not authorized by Congress. However, both Presidents Bush and Clinton adopted similar Executive Orders.

The House of Representatives in 1994 adopted regulatory impact assessment legislation, HR9, which would require RIAs for major rules imposing public and private costs of at least $25 million annually ($50 million for small business costs) and on Superfund cleanups costing over $5 million. The RIA requirements would apply to all new federal environmental, health, and safety regulations, including new regulations under existing laws.

Risk assessment refers to identifying how many lives will be saved or other benefits achieved by the proposed regulation. Benefit-cost analysis in the RIA context refers to determining how much it will cost per life saved under the proposed regulation. The RIA process is intended to identify proposed regulations with "high" cost-benefit ratios.

HR9 stalled in the Senate and is unlikely to be adopted in 1997. Nonetheless, the bill raises fundamental environmental policy issues that relate directly to ground water valuation. The overall effect of RIA requirements would be to discourage rules for which a positive benefit-cost analysis cannot be generated or is marginal.

Unfunded Mandate Act of 1995

While general RIA legislation has not been enacted, elements of the HR9 RIA process were incorporated in the Unfunded Mandates Act of 1995. Federal programs often require state and/or local governments to bear a significant portion of program implementation costs. This practice is now referred to by critics as creating an "unfunded federal mandate." An example of an unfunded mandate is the Safe Drinking Water Act. Communities whose water supplies violate SDWA standards must either treat their water to EPA levels or obtain a new water supply. No federal funding is available to meet these costs, hence the SDWA imposes an "unfunded mandate."

President Clinton has signed legislation requiring that unfunded (or underfunded) federal mandates be identified before they are adopted by Congress (109 Stat. 48, March 22, 1995; Fort, 1995). Unfunded mandates costing states and/or local governments at least $50 million annually are subject to a separate vote on establishing the unfunded mandate if a vote is requested. Cost estimates are prepared by the Congressional Budget Office. When faced with an unfunded mandate "point of order," Congress has several options: (1) provide the funding, (2) delete the mandate, (3) reduce the mandate to fit the funding, or (4) approve the unfunded mandate.

The Unfunded Mandates Act also requires federal agencies to prepare assessments of any proposed regulations imposing more than $100 million in combined compliance costs for states, local governments, and the private sector. Federal agencies are prohibited from issuing regulations containing federal mandates that do not (1) employ either the least costly or most cost-effective method or (2) do not have the least burdensome effect on the governments or private sector, unless the agency also publishes an explanation of why the more costly or more burdensome method was adopted. These regulatory review requirements are subject to judicial review.

The act is not retroactive but would apply when existing laws are reenacted. Most federal environmental, health, and safety programs contain program requirements that may be considered unfunded mandates. Accurate information regarding ground water values would make unfunded mandate regulatory reviews better relative to evaluation of the economic and environmental trade-offs involved in ground water protection policies.

LEGAL ISSUES IN REDEFINING GROUND WATER RIGHTS

Clearly, how private rights for ground water use are legally specified affects the value of that property right and the associated land. Changes in ground water right specifications may threaten property values and may likely be resisted by ground water users. Ground water users traditionally have resisted legal changes that affect the quantity they can pump, even though these restrictions (often little more than required improvements in ground water use efficiencies) may significantly extend aquifer life. Political barriers to reforming ground water law often are significant, and reform efforts unfortunately are most successful in response to a perceived ground water crisis.

One often criticized aspect of ground water rights is uncertainty. However, the lack of specificity regarding ground water rights most often reflects a lack of scarcity. As ground water becomes more scarce, its rights to use will become more sharply defined through litigation and/or legislation.

Takings and Property Rights

Another legal issue is that of takings. Often when property rights are modified, for example, through environmental regulation, the takings issue is raised. Simply stated, the U.S. Constitution requires that if government regulation of a property interest amounts to a complete destruction of that property interest, the property owner must be compensated or the regulation not implemented. Traditionally, states have enjoyed broad discretion to regulate property rights in the public interest. Recently, however, federal courts have begun to consider with more favor landowner complaints that some government regulations constitute a regulatory taking or partial taking of private property. While takings jurisprudence is hardly settled, this trend in takings law may make government agencies less inclined to take aggressive actions to protect ground water and more amenable to negotiating policies with all stakeholders, rather than simply imposing predetermined ground water policies. However, ground water regulations are routinely upheld in court, and efforts to defeat ground water regulations as takings will likely fail.

REDUCING RISK AND VALUING GROUND WATER

As noted in Chapter 1, more than half of the United States depends upon ground water as a source of drinking water, and the nation's reliance on ground water for drinking purposes is increasing. Moreover, ground water supplies contribute 30 to 40 percent of the flow of the nation's streams, an important ecological factor. The base flow provided by ground water contributes significantly to stream flow during periods of low precipitation. During drought periods stream flow may be entirely or largely derived from ground water discharge. Thus even the significant 30 to 40 percent contribution of ground water to total stream flow understates the significance of base flow. During droughts, most or all of the water in a stream is likely to derive from ground water discharge. Moreover, ground water plays a crucial ecological role in sustaining wetlands (NRC, 1995)

Despite the obvious strategic importance of ground water to human health and the environment, the EPA Science Advisory Board (SAB), in its influential report *Reducing Risk*, somewhat surprisingly concluded that ground water pollution represented a relatively low risk to natural ecology and human welfare (U.S. EPA, 1990, page 13). Despite acknowledgment that the natural recovery time from ground water contamination was centuries (U.S. EPA, 1990), the SAB seemed unaware of the significance of ground water both as the nation's primary drinking water source and as a major source of stream flow. The Board's conclusions are inconsistent with the drinking water and ecological services provided by ground water.

This study, on the other hand, finds that ground water is a resource critical to the nation's future, both as the our primary source of drinking water and as a significant source of stream flow. Because of these crucial human welfare and ecological functions performed by ground water, and because of the acknowledged difficulty of ground water remediation, protecting and maintaining the quantity and quality of the nation's ground water supplies must receive a high priority.

RESEARCH NEEDS

The published literature related to ground water valuation is limited. The studies on the use of valuation methods mentioned in Chapter 4 are characterized by a limited focus and budget (the one exception being the national study by McClelland et al., 1992). None of the studies completely address ground water's TEV; rather, attention was given to the valuation of extractive services or other specific components within the TEV. And since most of the case studies were conducted in academia, funding was limited. Further, most of the published literature and case studies were authored by economists and if an interdisciplinary approach was utilized, it was not documented.

Environmental economics, particularly as related to ground water valuation, must be seen as an emerging profession. It must be based on an interdisciplinary approach when applied to ground water development, protection, or remediation projects. However, several barriers to such interdisciplinary efforts exist. Two of particular significance are the absence of any standardized terminology to facilitate verbal and written communications, and internal organizational structures within regulatory, governmental, the private, and consulting sectors that tend to isolate economists into small groups with specific and narrow functions. Therefore, there is a need for capacity building related to professionals who can address ground water valuation. In other words, ground water professionals who are not trained economists should learn enough economics to be able to understand and use another professional's ground water valuation. As additional information becomes available, it should be transferred to practitioners.

At present, no integrated and comprehensive research program on ground water valuation exists. For example, comparative studies of common ground water management decisions that could be aided by valuation have not been completed. Likewise, for a given type of decision, such as whether or not to initiate a remediation program, a range of solutions should be explored.

Accordingly, a fundamental recommendation resulting from this NRC study is that the EPA, along with the National Science Foundation, DOE, DOD, and other agencies as appropriate, should plan and implement an integrated and comprehensive research program focused on ground water valuation. Following are suggestions of ways that this research need could be addressed:

- Pertinent federal agencies such as EPA, DOE, and DOD should jointly sponsor a series of research and case studies that examine: (1) a range of ground water-related decisions that could be facilitated by valuation information; and (2) a range of hydrogeological and contamination conditions that could be used to reflect both go/no go and prioritization decisions related to remediation. Such studies should include the integration of remedial investigation/ feasibility studies, risk (or endangerment) assessments, impact studies, and valuation information. An additional research question that could be explored based on both existing studies and the proposed research and case studies is exactly how the valuation information is used in each decision context; that is, was valuation the sole factor, an equal consideration, or a supplemental item of information?
- The appropriate composition of interdisciplinary teams to conduct ground

water valuation research needs to be explored. Participants should include environmental economists, ground water scientists and engineers, and political scientists. The roles of such disciplines and the constraints related to interdisciplinary endeavors need to be studied. Also, the public's perception of the use of valuation methods and its understanding of valuation study results should be gauged.

- Attention should be given to developing and improving methods for quantifying the value of ecological services and for determining existence and bequest values for ground water resources. These topics have so far received minimal attention. Of particular significance is both the recognition of ecological services and the development of integrated methods to quantify such services. Efforts should also be focused on developing TEV information for ground water resources.
- Technical and economic uncertainties must be recognized in efforts to develop site-specific information regarding the benefits and costs of decisions related to development, protection, and/or remediation of ground water. For example, in relation to remediation decisions, the stochastic modeling of the contamination problem and the potential effectiveness of cleanup measures should be used to develop ranges of information that can be viewed together as an indicator of "sensitivity analysis." The possible influences of uncertainties and nondelineated costs and benefits (or ground water services) should also be considered in a qualitative manner.
- Finally, because we are only recently recognizing the importance of valuing ground water, there are extensive educational and technology transfer needs for ground water professionals, environmental economists, regulators, and ground water managers. As noted earlier, the planning and conduct of such valuation studies requires interdisciplinary involvement. While the leadership should come from EPA, other federal, state, and local agencies, along with appropriate professional groups such as the Association of Ground Water Scientists and Engineers (AGWSE), American Geophysical Union (AGU), American Water Resource Association (AWRA), and others, should jointly participate in technical and policy conferences and in generating a body of published literature on the value of ground water and the advantages and limitations of valuation methods. Handbooks related to planning such studies should be developed. Such efforts would help to increase awareness of ground water's value and to demonstrate how valuation information could improve decision-making.

It is recognized that many of these stated research needs are broad. Such needs could be made more specific upon the development of a research strategy focused on ground water valuation. The strategy which should be spearheaded by EPA, should include an overall goal, specific objectives, delineation of multiple agency involvements (e.g., EPA, USGS, Corps of Engineers, NOAA, and nongovernmental organizations), specific problem statements on needed research topics, and budgetary requirements and time schedule.

RECOMMENDATIONS

These institutional considerations, suggest several areas of governmental action:

- **Federal, state, and local agencies should give consideration to the TEV of ground water in their deliberations on new or amended legislation or regulations related to ground water management.**
- **States should consider the authorization and promotion of water marketing, including transfer of ground water rights when appropriate. Although a transition to a market that adequately captures the full value of the resource may be difficult, water markets provide flexibility in water use and more efficient allocation of water among uses. Water markets also provide real world prices of water for current use values, and their prices aid decision-makers in valuing ground water. Helping to drive water marketing is the fact that the importance of ground water has changed in the context of conjunctive use. Recharge of surface water and effluent to replenish ground water is now common in southern California.**
- **States should be encouraged to develop clear and enforceable rights to ground water where such rights are either lacking or absent. A system of clear and enforceable extractive rights to ground water is prerequisite to economically efficient use of that water. Without such rights users will not have the incentive to value ground water correctly either now or in the future.**
- **Because of many uncertainties related to ground water valuation as demonstrated in both the methods chapter (Chapter 4) and this chapter, EPA and other pertinent agencies should plan and implement an integrated comprehensive research program on ground water valuation. Federal agencies should conduct research and develop case studies in ground water valuation that includes a range of environmental conditions and economic circumstances. In addition, federal, state, and local agencies should develop valuation methods that quantify ecological services and bequest and existence values. Such research will help states manage and protect their ground water resources and could help to demonstrate improvements in decision-making that would occur with valuation information.**

REFERENCES

Aiken, J. D. 1982. Ground water mining law and policy. Colorado Law Review 53(3):505-528.

Beck, R. E. 1991. Environmental controls. Waters and Water Rights 5: Chapters 52-57. Charlottesville, Va.: Mitchie Company.

Brown, F. L., S. C. Nunn, J. W. Shomaker, and G. Woodard. 1996. The value of water: A report to the city of Albuquerque in response RFP95-010-SV. Albuquerque, NM: City of Albuquerque.

Copple, R. F. 1995. NOAA's latest attempt at natural resource damages regulations: Simpler... but better? Environmental Law Reporter 25:10671.

Fort, D. D. 1995. The unfunded mandate reform act of 1995: Where will the new federalism take environmental policy? Natural Resources Journal 35(3):727-730.

McClelland, G. H., W. D. Schulze, J. K. Lazo, D. M. Waldman, J. K. Doyle, S. R. Elliott, and J. R. Irwin. 1992. Methods for Measuring Non-Use Values: A Contingent Valuation Study of Groundwater Cleanup. Boulder: University of Colorado.

Murphy, E. F. 1991. Quantitative Ground Water Law. Waters and Water Rights 3: Chapters 18-24. Charlottesville, Va.: Mitchie Company.

National Research Council. 1992. Water Transfers in the West. Washington, D.C.: National Academy Press.

National Research Council. 1994. Ranking Hazardous Waste Sites for Remedial Action. Washington, D.C.: National Academy Press.

National Research Council. 1995. Wetlands: Characteristics and Boundaries. Washington, D.C.: National Academy Press.

Olson, E. D. 1989. Natural resources damages in the wake of the Ohio and Colorado decisions: Where do we go from here? Environmental Law Reporter 19:10551-10557.

Seiver, D. W. 1996. Law of chemical regulation and hazardous waste. 1:§6.06[2]. Deerfield, Ill.: Clark Boardman Callaghan.

Tarlock, A. D. 1995. Law of Water Rights and Resources. Chapters 4, 6. Deerfield, Ill.: Clark Boardman Callaghan.

U.S. Environmental Protection Agency Science Advisory Board. 1990. Reducing Risk: Setting Priorities and Strategies for Environmental Protection. Washington, D.C.: Environmental Protection Agency.

6

Case Studies

This report has emphasized the importance of valuing ground water resources and suggested a framework and valuation methods that could be used to quantify the economic values associated with a suite of ground water services. This chapter provides brief descriptions of seven existing situations that highlight the importance of valuing ground water resources. These case studies also include some information on applicable valuation methods. The chapter offers some insight into the difficulties that water managers (and policy-makers in general) face in attempting to translate recommendations regarding valuation methods into usable estimates of ground water values. Such difficulties can derive from institutional constraints or conflicts in specific locales; political considerations; terminology and conceptual problems related to communicating information; and uncertainties associated with technical analyses, determination of effects, and economic assumptions.

These site-specific studies are brief and are not intended to offer solutions for any other case. Instead, these examples demonstrate that valuing ground water resources is not a recipe that can simply be followed at any site. The planning and implementation of economic valuation studies requires the interdisciplinary efforts of economists, engineers, scientists, and policy-makers. These studies show that although a complete accounting for all components of the TEV of ground water is often impossible to obtain, quantifying some components can provide information to improve decision-making and increase the efficiency of the use of scarce ground water resources. Table 6.1 summarizes the theme of each case study.

The first case study illustrates the link between surface water use and the

TABLE 6.1 Comparative Information on Seven Case Studies

Case Study	Theme	Comments
Treasure Valley, Oregon	Linkage between surface water usage for agriculture and the value of ground water services whose quantity and quality may be influenced by agricultural practices.	Illustrates importance of ground water valuation in designing allocative and management policies for the conjunctive use of surface and ground water.
Laurel Ridge, Pennsylvania	Use conflicts that may arise among local governmental agencies coordinating various combined uses of surface and ground water.	Illustrates need for systems approach in defining hydrogeology, surface and ground water resources, and competing uses within a multi-institutional framework in a local geographical area.
Albuquerque, New Mexico	Development of long-term water use strategy for a city that relies on ground water for its water supply; also includes buffer value of ground water.	Includes information on both use and nonuse value of ground water and how this can be incorporated in long-term water supply planning in an area where ground water mining occurs.
Arvin-Edison, California	Buffer value of ground water in an area subject to periodic droughts.	Demonstrates buffer benefits of a ground water resource in an agricultural area.
Orange County, California	Use of ground water recharge in a coastal area to avert sea water intrusion in a viable ground water basin.	Addresses the value of ground water in storage as a deterrent to sea water intrusion.
Woburn, Massachusetts	Incorporation of the value of ground water in deciding on remediation for a Superfund site.	Illustrates numerous uncertainties associated with local hydrogeological conditions, pollutant transport, the effectiveness of remediation strategies, and direct and perceived health consequences of drinking contaminated ground water.
Tucson, Arizona	Planning for application of valuation framework to decisions for meeting water demand; options addressed are ground water recharge and/or surface water treatment.	Illustrates the variety of considerations associated with a ground water valuation study, including the need to incorporate engineering estimates along with valuation methods; also focuses attention on the importance of substitute water supplies.

quantity and quality of ground water in the Treasure Valley area of eastern Oregon and southwestern Idaho. In this setting, which is typical of many areas in the West, agriculture relies primarily on surface water supplies, and ground water is used mainly for human and industrial consumption. The presence of pollutants from agriculture in an aquifer reduces the value of the ground water for human consumption and poses challenges for water resource managers. Without estimates of the value of the services associated with unpolluted ground water, managers may design allocation and management policies that could lead to suboptimal use of both the scarce ground water and the surface water supplies.

The Laurel Ridge, Pennsylvania, case study is an example of competing uses of an aquifer and the interplay between ground water and surface water supplies. In this area the user conflicts are between development (mining) and tourism and among the many fragmented local governments whose jurisdictions overlay the watershed. Economic valuation is a crucial component to achieving a more systematic approach to planning in this watershed.

The next two case studies deal with the buffer value of ground water. In Albuquerque, New Mexico, ground water is the primary source for municipal water supply, although the city also has rights to surface water from nearby rivers. Recent concerns with both the size of the aquifer and increased population growth along with ground water mining have initiated a series of engineering and economic studies to assess the long-term strategies for water use. This example provides concrete evidence of the role that economic values can play in formulating policy alternatives for water use management.

The Arvin-Edison Water Storage District in southern California is another example of a buffer value success story, where the surplus water from wet years is being used to recharge the aquifer. This water management system in the Bakersfield area has been in place for nearly 30 years and by some estimates has generated millions of dollars in net returns to agricultural interests that would have been foregone during critically dry years.

The second California example deals with the issue of irreversibilities associated with the intrusion of seawater in the ground water basin underlying Orange County, in southern California. Loss of the basin to sea water intrusion would require the Orange County Water District to rely more heavily on imported water and would preclude the use of the aquifer for water storage. Knowing the value of the ground water was clearly an important component in the decision to construct and operate Water Factory 21 (an advanced wastewater treatment plant) and two water injection projects. Combinations of imported water and highly treated municipal wastewater are recharged as a barrier to sea water intrusion.

The sixth case study, a Superfund example, illustrates the importance of ground water valuation to federal regulations regarding remediation of contaminated aquifers. Policy decisions on the extent to which ground water remediation should be pursued need to be based on a careful assessment of the costs and benefits of proposed actions. The benefits of restoring the quality of a contaminated aquifer will be reflected in the potential gains or value of improvements to the ground water resource and will be site-specific.

The empirical findings of this Woburn, Massachusetts, case study refute conventional wisdom concerning the economic efficiency of ground water remediation at Superfund sites for the sole purpose of restoring drinking water supplies (i.e., that the costs of remediation far outweigh the benefits). In some cases ground water remediation can be the efficient alternative; it should not be dismissed without conducting a cost-benefit analysis. This case study also highlights the complexities involved in conducting an empirical analysis of the value of restoring ground water resources and the impacts of uncertainties in the economic and physical dimensions, and in potential health consequences, and the public response to ground water usage.

The final case study concerns the potential application of the valuation framework described in Chapter 3 and some valuation methods described in Chapter 4. Options in this Tucson, Arizona, case include ground water recharge using Central Arizona Project (CAP) water or treatment of CAP water prior to usage. This study provides information on the types of methods that could be used to value a complete suite of ground water services for both options.

CHALLENGES IN WATER QUALITY MANAGEMENT
Treasure Valley, Oregon

Background

The Treasure Valley of eastern Oregon and southwestern Idaho is high desert (10 inches of precipitation on average per year) that is intensively irrigated using surface water from the Owyhee, Malheur, and Snake Rivers. All the water of the Owyhee and Malheur Rivers (tributaries of the Snake River) is diverted to irrigation. Stream flow below the diversions is maintained by irrigation return flows and recharge from a shallow aquifer supported in part by irrigation recharge (Gannett, 1990).

Crop agriculture in the area is characterized by a range of high valued crops including potatoes, sugar beets, and onions, as well as cereal grains and hay. In the Oregon portion of the valley, approximately 180,000 acres are in irrigated crop production (Schneider, 1992). The primary source of water irrigation is from federal (U.S. Bureau of Reclamation) reservoirs and distribution systems. In terms of total agricultural sales, animal agriculture (cattle and dairy) accounts for 36 percent of sales, onions 25 percent, potatoes 11 percent, sugar beets 9 percent, cereal grains 9 percent, and the remaining crops 10 percent.

Ground water is used largely for industrial or human consumption. Between 1983 and 1986, the Oregon Department of Environmental Quality (ODEQ) tested water wells in the study area. Elevated nitrate levels were found in 67 percent of the wells tested; 35 percent of the wells exceeded the federal drinking water standard for public water supplies of 10 mg/l. In 1989 ODEQ declared Malheur County a ground water management area and ordered that ground water nitrate levels be 7 mg/l or less by the year 2000. The ODEQ and local water quality management groups have identified agriculture as the primary contributor to ground water nitrates. Pesticides (dacthal) associated with onion production have also been found in test wells.

Valuation/Management Issues

The geohydrological link between surface water applications and ground water quality and quantity found in Treasure Valley is typical of many ground water situations in the West. Specifically, percolation of irrigation water serves to recharge the ground water aquifer (and in this case surface water percolation augments the natural flow in the aquifer). This ground water recharge/augmentation process serves a number of beneficial purposes. For example, recharge increases seepage from the aquifer into lowlands, creating wetlands for wildlife. Irrigation returns, whether through surface runoff or through eventual seepage of ground water to the Snake River and its tributaries, helps to stabilize stream flows. However, unwelcome consequences may accompany this recharge, including the elevated levels of agricultural pollutants of concern to the ODEQ.

Ground water is the primary source of water for household and industrial uses around Ontario, Oregon, located near the center of the valley (Gannett, 1990). The presence of pollutants from agriculture, with associated health concerns, reduces the value of water for human consumption. Pollutants in ground water also degrade water quality in streams, with possible adverse consequences for fish and wildlife. Given present concerns about endangered salmon fisheries in the Snake River (the U.S. Fish and Wildlife Service have listed Snake River sockeye and chinook salmon as endangered), water quality has assumed increased importance.

A number of strategies to reduce the amount of agricultural effluents reaching the aquifer have been proposed. A feature common to most strategies is "better" irrigation water management, which implies less total water application per acre and hence less deep percolation. Such practices, however, also reduce the volume of water moving into the aquifer. This in turn affects the volume of seepage into wetlands and return flows to rivers. Further, if irrigation water "saved" by improved irrigation management is used to expand irrigated acreage, the total return flow and hence stream flow may be markedly reduced. Reduction in stream flow and wetlands will exacerbate some wildlife problems.

Assessment of the Value of Ground Water

The interplay of surface water use, ground water quality, and, ultimately, stream flow, creates challenges for public water resource managers as they try to achieve multiple objectives. Institutional constraints, including the nature of water rights (prior appropriation doctrine) and below-cost pricing of water in public supply projects, further complicates water management. A plan that achieved optimal use across all water resources in the basin would likely vary dramatically from the use pattern typically observed in such settings. Assessment of the values from one type of water resource, such as ground water, in isolation will lead to suboptimal resource use.

To date, the benefits of ground water quality or ground water services in general have not been estimated for this area because of the focus on human health issues. Specifically, federal and state regulations require that water quality in the aquifer be brought into compliance with state water-quality standards. Economic analysis has been limited to assessment of the consequences to farmers of meeting the standards (Fleming et al., 1995; Connor et al., 1995). An understanding of the values of ground water could aid in comprehensive management of water.

Against this backdrop of complex geohydrologic linkages, institutional constraints, and a regulatory mandate to improve water quality, it is instructive to consider whether the valuation techniques discussed in Chapter 3 can be used to estimate the value (benefits) of the ground water services provided here. The answer is a qualified yes. For example, the value of unpolluted ground water for household uses can be estimated through expenditures on averting behavior, such as purchase of bottled water or purification systems. Values of stream flow for recreational fishing can be estimated through travel cost procedures. Direct elicitation of nonuse values to maintain or enhance a species (e.g., existence values) could be estimated by the contingent valuation method, although the costs of performing defensible CVM surveys are quite high. Similarly, TCM or CVM can be used to determine the value of ground water recharge of wetlands for both use and nonuse services the wetlands provide. A compilation of these use and nonuse values would supply information on the trade-offs between management goals across water users, including protection of ground water services.

References

Connor, J. D., G. M. Perry, and R. M. Adams. 1995. Cost-effective abatement of multiple production externalities. Water Resources Research 31:1789-1796.

Fleming, R. A., R. M. Adams, and C. S. Kim. 1995. Regulating groundwater pollution: Effects of geophysical response assumptions on economic efficiency. Water Resources Research 31:1069-1076.

Gannett, M. W. 1990. Hydrogeology of the Ontario Area, Malheur County, Oregon. Ground water Report 34. Salem: Oregon Department of Water Resources.

Schneider, G. 1992. Malheur County Agriculture. Ontario: Oregon State University Extension Service.

COMPETING USES OF AN AQUIFER
Laurel Ridge, Pennsylvania*

Background

Laurel Ridge covers 330 square miles in southwestern Pennsylvania. The generally forested, mountainous topography forms a distinct break with the surrounding plateau lowlands. An estimated 15 million tourists visit Laurel Ridge each year. Recreational activities such as hunting, fishing, boating, and skiing are supported by abundant, clean water and large holdings of public land (41 percent of the area). The dominant land uses of Laurel Ridge, such as recreation, water supply, wildlife habitat, and forestry, contrast with those of the peripheral lowlands, which are largely devoted to agricultural pursuits and coal mining. While tourism is an invaluable resource to communities within the area, high rates of unemployment and slow growth in other economic sectors persist. This area also has the highest acidic deposition in Pennsylvania. The Allegheny and Pottsville rock units are influenced by acid deposition and yield ground water high in hydrogen ion concentration and dissolved aluminum. Buffering from the Mauch Chunk/Burgoon aquifer and its discharges into area streams help support aquatic life (Beck et al., 1975).

Pennsylvania government is fragmented. With over 2,500 minor civil divisions, the state ranks second in the nation in terms of the number of local government divisions. The Laurel Ridge region reflects this fragmentation: parts of four counties (Somerset, Cambria, Fayette, and Westmoreland) come together along the historic ridge-line boundary; within these counties, 22 townships and two boroughs form an intricate web of administrative jurisdictions. Thus the natural resources of the Laurel Ridge are not managed as a cohesive region.

Ecosystem Characteristics

The Mauch Chunk/Burgoon aquifer is the only source of high-quality ground water in the Laurel Ridge. It supplies most of the total public and domestic water supply and provides base flow to many of the region's exceptional surface waters. Compliance with the 1986 amendments to the federal Safe Drinking Water Act requires that all surface water used as drinking water for public water systems be filtered. From 1990 to 1995, some 30 high-yield municipal water wells were drilled in the area. The aquifer supplies high-quality upland streams through

---

*William Delavan, Graduate Research Assistant, and Charles Abdalla, Associate Professor, Department of Agricultural Economics and Rural Sociology, Pennsylvania State University, prepared this case study. Information in this case study was obtained through personal interviews with faculty at Pennsylvania State University and with Pennsylvania Department of Environmental Protection staff.

artesian head-water springs. The effect of this development on streams has raised concerns about both the quantity of water withdrawn and the impact on water quality. Specifically, changes in withdrawal patterns have threatened aquatic environments that support fish and other organisms. Water quality is further affected by a combination of geographic and geologic factors that create in one of the highest rain acidities in the country.

Water Users and Use Conflicts

The rapid development of the aquifer, the lack of rules to allocate ground water among competing uses, and, in most cases, the absence of local water management and planning has led to a situation where it seems the person with the biggest pump or deepest well wins. Currently, there is little economic incentive for users to conserve. Since regulation is likely to occur in the future, users who establish an early claim to the resource stand to win by drilling before regulations are developed and carried out.

There are several conflicting interests. The legacy of coal mines is prevalent throughout Pennsylvania. On both sides of the ridge in the lowlands there is degradation from coal mining; the aquifer is thus threatened on its boundaries. Assigning responsibility for past damage from coal mining is problematic from both a political and economic standpoint.

The region is home to two destination resorts whose ground water withdrawals are generally substantial from late November to early April. The resorts have recently established golf courses that have increased off season withdrawals. A rise in the number of second homes on the ridge ("suburbanization") has multiplied water demand. The impacts of the resorts' usage are not well understood. Some parties argue that efforts to recycle runoff and sewage serve to increase or maintain ground water levels by replacing water on the ridge, in effect performing an environmental service. Others deny this claim and fear that the resorts' usage threatens water quality down slope. Furthermore, the ground water pumping may move waters out of areas favorable toward fish stocks and recharges areas unfavorable to fish stocks, compromising wildlife habitat.

The resorts have a significant economic impact in providing employment as well as an influx of tourist dollars. Are the benefits of development greater than the costs in terms of resource degradation and other foregone opportunities? If, on the other hand, development inspires resource decisions that have high costs or are irreversible, such as ground water contamination by toxics, the sustainability of the local economy and its ecological systems is called into doubt. If, on the other hand, restrictive regulations or the absence of a plan to provide for long-term water and sewer requirements inhibits development, then attempts to attract new industry and lower the unemployment rate will be stymied.

Issues Related to Economic Valuation

Efforts to understand the physical systems of the watershed must be combined with equal efforts to measure how people value these systems. Policy-makers must address four issues:

- How much water is safely available from this aquifer system, and are there areas where the aquifer is potentially overdeveloped?
- What is the impact of ground water withdrawals on the quality and base flow of the upland surface water systems fed by Laurel Hill Spring?
- How can economic values for different uses be measured so that decision-makers may adequately take into account competing uses?
- Can effective watershed management increase the potential for optimizing the different uses? What is the best way to develop institutions to help carry out comprehensive planning?

Economic Values and Decision-Making

Regional watershed organizations have stepped up to meet these challenges, but their efforts may be insufficient to educate the public and measure and map resources. Even armed with accurate knowledge of ground water functions, policy-makers face complex decisions. The fragmented nature of municipal government in Pennsylvania poses serious challenges to intercommunity communication and cooperation, challenges that may be overcome only by a more systematic watershed approach to planning and policy implementation.

In April 1992, the Laurel Ridge Forum was created in recognition of the region's vast public holdings, outstanding natural resources, and recreational opportunities. Composed of members from state and local government, business, and water suppliers, the forum focuses on future development in the area. Water rights conflicts between residents and second-home owners are at the center of the development debate. Research is beginning to define the physical impact of recreational uses and the extent of past degradation from coal and limestone mining, brine disposal, and road salting. Economic valuation is necessary to interpret how different members of the community value these environmental changes. New or different institutional arrangements among the layers of government could facilitate the comprehensive and systematic management of natural resources. The Laurel Ridge Forum's Coordinated Resource Management Plan (CRMP) attempts to deal with governmental fragmentation.

Decision-makers must identify and study alternative policies for effectively managing these water resources. It might make sense, for example, to manage the watershed as well as the basin as a whole. As research defines the aquifer's physical limits and capabilities, stakeholders and decision-makers must continue to ask questions about the economic value of ground water. Specifically, they need to better understand and quantify the economic benefits of protecting the aquifer from depletion or degradation.

The Laurel Ridge area offers a unique and challenging context for ground water valuation. Rapid development and competing interests have brought the water issue to the forefront, forcing increased efforts to understand and measure water resources and begin constructive public debate. Economic values will allow local officials to make more informed decisions relative to resource use by helping them gauge the community's values regarding water resources and the tradeoffs between protecting these resources and economic development. Economic valuation coupled with a comprehensive systems approach to the watershed should guide decision-makers toward effective water resource management choices.

Reference

Beck, M, G. Cannelos, J. Clark, W. Curry, and C. Loehr. 1975. The Laural Hill Study: An Application of the Public Trust Doctrine to Pennsylvania Land Use Planning in an Area of Critical State and Local Concern. Department of Landscape Architecture and Regional Planning. Philadelphia: University of Pennsylvania.

THE BUFFER VALUE OF GROUND WATER
Albuquerque, New Mexico

Background

The city of Albuquerque, New Mexico, like many other rapidly growing metropolitan areas in the arid Southwest, draws much of its municipal water supply from ground water. Unlike most other cities, however, Albuquerque does have rights to surface water supplies from the nearby middle Rio Grande and to waters from the Colorado River basin (San Juan and Chamba Rivers) that are diverted to the Rio Grande basin. The city's historical reliance on pumping ground water in lieu of accessing available surface water reflects a mix of geohydrological, institutional, and cultural forces. These forces are changing and call into question the economic and physical sustainability of Albuquerque's water use patterns.

In response to concerns over the long-term viability of ground water pumping, the city commissioned a series of engineering and economic valuation studies to assist managers in developing sustainable management strategies (CH2M-Hill, 1995; Boyle Engineering, 1995; Brown et al., 1995). In addition, other agencies involved in water issues in the area have issued or commissioned studies pertaining to water (Middle Rio Grande Conservancy District, 1993; EcoNorthwest, 1996). Albuquerque's strategies for water use, as described in these studies, provide examples of the role economic values can play in assisting policy formation.

The middle Rio Grande valley has been inhabited and intensively farmed by Native Americans for at least 500 years. In addition to providing a stable water supply for irrigation, the riparian, tree-lined areas, or bosque, along the River were important to Native Americans for wood for fuel and shelter as well as cultural and spiritual purposes. Hence, communities (pueblos) sprang up at points on or near the River and its tributaries. Europeans were also attracted to the riverine environment of the Rio Grande valley and established settlements on the sites of present-day cities such as Albuquerque.

As settlement progressed and the region grew, residents encountered new water issues. Competition among states (Colorado, New Mexico, and Texas) and between the United States and Mexico for the scarce surface water supplies of the basin resulted in a series of compacts and agreements allocating water among the parties. Albuquerque was given rights to 48,000 acre-feet of water from the Rio Grande and 22,000 acre-feet of imported water from the Colorado River basin. Total surface allocations in the middle Rio Grande basin exceed 350,000 acre-feet; they are used primarily for irrigated agriculture.

While agriculture relies heavily on surface water, the settlements in the valley, including Albuquerque, have relied heavily on ground water to meet the needs of the increasing population. Albuquerque sank deep wells as early as 1910 to secure municipal water. This use of ground water was motivated in part by the high quality of ground water, the steady supply (even in years of drought) and the belief that the aquifer supply was large and recharge rapid. Rapid recharge of the aquifer from the River led city water managers to believe that they were simply pumping their surface water allocation, albeit with a slight lag time.

Present Situation

Recent geohydrological information that challenges past assumptions, increased competition for water, continuing population pressures, and concerns over the environmental health of riverine habitat in the middle Rio Grande valley cast doubt on the wisdom of Albuquerque's reliance on ground water. Perhaps the most important development was a 1993 U.S. Geological Survey study that revealed that the aquifer was not as large as originally believed nor is recharge (from surface flows) as rapid as assumed. This meant that Albuquerque was not using its surface water supplies but was instead mining or overdrafting its ground water. Inventory information also suggested that if Albuquerque continued to rely on the aquifer to meet its urban needs, the aquifer would be economically exhausted by 2060. During this same time period, the U.S. Fish and Wildlife Service (USFWS) listed the Rio Grande silvery minnow, found in the middle Rio Grande, as an endangered species. To ensure survival, the USFWS proposed increases in instream flows and protection of riparian habitat. Meeting these instream and other habitat needs implies changes in water use patterns.

Once city water managers understood that Albuquerque was mining ground water and not using its surface water supplies, they reexamined the long-term water management strategy. The city's failure to use its surface water supplies meant that someone else had been using those supplies. The significance of the use issue is contained in western water law; specifically, western water law requires that users demonstrate a beneficial use of water within a specific time period. While cities may be treated differently from other (private) users of water, increased competition for this water places pressure on the city to begin actively using its allocation. However, the total water allocation (of 70,000 acre-feet) is not adequate to meet future needs. Thus some combination of policy options, including securing alternative surface water supplies, most likely from agriculture, and increases in urban rates to reduce consumption, will be needed if the city wishes to develop a sustainable aquifer management policy.

Valuation Issues

The situation in Albuquerque is similar to that in many other cities in arid regions of the West. Historical preference for use of ground water in meeting urban needs reflects some of the advantages ground water provides, including stability of supply, high water quality (no treatment of ground water is required in Albuquerque), and ease of access (no collection and transport system is required, as in the case of most surface water supplies). The value of these advantages is typically not reflected in the "price" of ground water (the "price" that cities charge consumers is usually set at the cost of pumping and distributing the water). A low price for ground water encourages higher use of the resource.

If ground water were not scarce (i.e., were available in unlimited quantities), then its price would simply be the cost of extraction. However, ground water, like surface water, *is* scarce; and when water is used in one setting, such as urban use, it is not available for another purpose, such as in riparian habitat enhancement. Ground water price should thus reflect not only extraction costs but also foregone benefits from its use in some other setting or in the same use but at some future time (its opportunity cost). Until recently, most cities did not include such values in the price of water.

In the presence of mining, as is occurring in Albuquerque, potential long term adverse effects jeopardize the flow of future services. The lost benefits (costs) from the reduced flow of these services should be reflected in water pricing. One of these effects is land subsidence (due to compaction of the pore spaces in the aquifer). Subsidence may lead to damages to buildings, roads, and other structures. Mining also affects water quality; water quality in aquifers tends to decline as pumping depth increases. Falling water levels in the aquifer also reduce the ability of the aquifer to maintain or support stream flows and maintain riparian zone health. Such drawdowns of water levels also increase pumping costs to all users. Eventually, mining eliminates the potential use of an aquifer as a buffer against drought. In arid regions, which are typically characterized by high annual variation in precipitation and surface water supplies, the use of ground water to meet needs during drought may be one of the most valuable ground water services.

Brown et al. (1995) examine a series of options or scenarios for the city to reduce aquifer use to a long-term, sustainable level by limiting use to periods of extended drought. Sustainability (to build up the aquifer to a level sufficient to provide a buffer against an extended drought) requires that the city live within its annual water budget as defined by renewable surface water supplies (again, except for periods of drought). The implications of ongoing use of the aquifer are short-term gains, accruing primarily to present users, with costs (of overdrafting) delayed to some future period (future generations), when the adverse effects described above would begin. Alternative strategies imply costs to present users but with potential long-term benefits. To weigh the benefits and costs of alternative actions requires the measurement of economic values, over time, for the array of services under the range of options available to the city.

In planning conjunctive management of the water resources of this region, policy-makers can benefit from an understanding of the value of water in its various uses. As they consider alternative water strategies, they should, as is practical, look at the full range of economic consequences associated with each alternative. The range of services affected by each option includes the potential for changes in both use and nonuse values. Use values in this case are as input in production (e.g., agriculture, manufacturing) and recreation; nonuse values are associated with the Rio Grande bosque, such as riparian habitat, endangered species, and aesthetic or visual services. Estimates of some use values in the region are discussed in Brown et al. (1995). Researchers have also measured nonuse (existence) values for provision of instream flows for preservation of the silvery minnow (Berrens et al., 1996). Thus information is available to assess some economic trade-offs involved in moving to a sustainable aquifer management policy.

The choice among alternative policies for ground water management should reflect, at a minimum, the opportunity costs of that decision (what is given up in selecting that option, or the benefits foregone from some other use of the water). A full accounting would include the willingness to pay for changes in services associated with each option (the maximum benefit or value associated with those services). The costs (lost benefits) are not likely to be spread uniformly or equally across affected parties. The political and judicial process can address some equity issues but typically does not reflect the interests of future generations. Only by achieving sustainability (by establishing a safe minimum reserve capacity in the aquifer) can the interests of future generations be guaranteed.

References

Berrens, R. P., P. Ganderton, and C. Silva. 1996. Valuing the protection of minimum instream flows in New Mexico. Journal of Agricultural and Resource Economics. In press.

Boyle Engineering. 1995. Water Conservation Rates and Strategies. Report prepared for Albuquerque, New Mexico.

Brown, F. L., S. C. Nunn, J. W. Shomaker, and G. Woodard. 1995. The Value of Water: A report submitted to the city of Albuquerque, New Mexico. Albuquerque, N.M.: City of Albuquerque.

CH2M-Hill. 1995. Albuquerque Water Resources Management Strategy: San Juan-Chama Options. Report prepared for the city of Albuquerque, New Mexico.

EcoNorthwest. 1996. The Potential Economic Consequences of Designating Critical Habitat for the Rio Grande Silver Minnow. Draft report prepared for the U.S. Fish and Wildlife Service, New Mexico field office.

Middle Rio Grande Conservancy District. 1993. Water Policy Plan; Working Document.

THE BUFFER VALUE OF GROUND WATER
Arvin-Edison Water Storage District, Southern California

The Arvin-Edison Water Storage District is located at the southern end of California's Central Valley, about 20 miles south of the community of Bakersfield. The district contains approximately 132,000 acres of highly productive agricultural land. The economy of the area is almost wholly dependent on agriculture, as there is little other industry. The value of agriculture in the district approaches $300 million annually, and land values range from $1,600 to $2,300 per acre. The principal crops include grapes, potatoes, truck crops, cotton, citrus, and deciduous fruit. Seventy-five percent of California's carrot acreage is found here. The climate is hot and arid, with average annual precipitation totaling only 8.2 inches. Almost all precipitation occurs between October and April. The sparseness and seasonality of precipitation means that irrigation is essential. On average growers apply 3 acre-feet of water per acre (Arvin-Edison Water Storage District, 1996).

Development of the area began after the turn of the century, and growers relied primarily on ground water supplemented by small and erratic flows from minor local streams. Most growers had their own wells and were responsible for providing their own supplies of irrigation water. As agriculture in the region grew, ground water extractions began to exceed rates of recharge and growers experienced declining ground water tables. Between 1950 and 1965, for example, water tables fell from an average depth of 250 feet to 450 feet. In 1965, average annual overdraft in the district totaled 200,000 acre-feet, which accounted for almost half the water applied districtwide. Continued overdrafting threatened the area's economic base.

Some years earlier local growers anticipated this situation and organized the Arvin-Edison Water Storage District to bring supplemental surface water supplies to the area to offset the overdraft. Beginning in 1966, Arvin-Edison received imported surface water from the Friant-Kern Canal, the southernmost component of California's Central Valley Project (CVP). The advent of significant surface water deliveries did not fully solve the area's water supply problems, however.

The district's water service contract called for annual importation of 40,000 acre-feet of firm (guaranteed) supply and up to 311,675 acre-feet of interruptible or nonfirm supply on an as-available basis. Although the district was subsequently able to increase the quantity of firm supply through an exchange arrangement, actual deliveries from 1966 to 1994 ranged from 30,000 acre-feet to almost 270,000 acre-feet. The problem lies with the significant portion of supply that is interruptible and therefore not available in years when precipitation is below average. This problem was resolved by percolating surplus supplies in wet years to recharge the underlying aquifer through the district's water-spreading facilities. Dry-year deficiencies were then offset by pumping previously percolated waters from the aquifer and delivering them to growers through the district's canal system (Vaux, 1986).

Over the period 1966-1994, more than 4 million acre-feet were imported to the district, 1 million of which were percolated to the underlying aquifer. Despite significant withdrawals to meet demands in the dry years of 1976-1977, 1982, and 1986-1992, net aquifer recharge has totaled 372,000 acre-feet and water table levels have stabilized. This has provided direct use benefits in the form of reduced pumping costs to approximately 20 percent of the growers in the district who are not connected to the distribution system and must continue to rely on direct ground water pumping. Perhaps more significant, the operation of Arvin-Edison's water supply system provides a clear illustration of the buffer value of ground water (Arvin-Edison Water Storage District, 1994).

In California less-than-average precipitation occurs with a frequency of about four years out of seven. To the extent that precipitation shortfalls are reflected in reductions in deliveries of surface water, ground water buffering values will be realized in each year that precipitation is less than average. The magnitude of the value will depend upon the degree to which surface water deliveries are deficient. In the critically dry years of 1977 and 1991, the surface water imports available to Arvin-Edison were 22 and 26 percent of average, respectively. Yet the district was able to make deliveries to water users that amounted to 85 and 90 percent of average annual deliveries, respectively. Rough calculations suggest that in 1991 more than 26,000 acres would have been fallowed had water stored in the aquifer not been available. Assuming typical cropping patterns and typical prices (in 1991 dollars) the gross value of production on this acreage exceeded $38 million. The returns to growers net of fixed and operating costs were almost $6 million (Arvin-Edison Water Storage District, 1994).

The use of ground water and aquifer storage capacity by the Arvin-Edison Water Storage District has yielded both direct use benefits and buffering benefits.

However, there are a number of issues that await resolution. The Metropolitan Water District of Southern California (MWD) is considering a long-term contract that would allow MWD to store water in the Arvin-Edison aquifer in wet years and withdraw it in drier years to meet urban and industrial demands in the Los Angeles area. Such a contract would increase the buffering value of the aquifer. Growers in the district face the issue of whether to renew contracts with the federal government for surface water supplies at prices reflecting full cost. These increased costs of surface water imports will need to be weighed against the present and future costs of pumping ground water as the sole source of irrigation water. It is clear that the availability of low-cost surface water that could be used for aquifer replenishment has sustained the agricultural economy of the Arvin-Edison District on a larger scale than would have been possible if ground water were the sole source of supply. The issue of whether the buffering value of imported supplies will be sufficient to offset potential increases in the cost of imported surface water remains to be resolved.

References

Arvin-Edison Water Storage District. 1994. The Arvin-Edison Water Storage District, Water Resources Management Program. Arvin, California.

Arvin-Edison Water Storage District. 1996. The Arvin-Edison Water Storage District, Water Resources Management Program. Arvin, California.

Vaux, H. J., Jr. 1986. Water scarcity and gains from trade in Kern County, California. Pp. 67-101 in Scare Water and Institutional Change, K. D. Frederick, ed. Washington D.C.: Resources for the Future.

THE VALUE OF AVERTING SEA WATER INTRUSION
Orange County, California

Background

The Orange County Water District (OCWD) operates and maintains a 15-million-gallon-per-day (mgd) reclamation sea water barrier project that protects a 350-square-mile ground water basin. OCWD constructed Water Factory 21 in 1973 for the purpose of protecting the quality of the county's extensive ground water resources by preventing sea water intrusion.

Loss of the basin beyond any possible use would require the district to rely on imported water for its entire water supply. However, this is not the only value of a ground water basin: the basin also provides storage and distribution, supplying water for peak and emergency use.

Sea Water Intrusion

Sea water intrusion occurs in ground water basins located along the coast. As overdrafting of a basin continues, the sea water front is drawn inland, threatening the ground water basin. Two fundamental conditions must exist before a ground water basin can be intruded by sea water. First, the water-bearing materials comprising the basin must be in hydraulic continuity with the ocean; second, the normal seaward ground water gradient must be reversed or at least too flat to counteract the greater density of sea water.

Sea Water Intrusion in Orange County

The largest body of ground water in Orange County is the coastal basin of the Santa Ana River, which yields most of the ground water produced in Orange County. The Santa Ana Gap is a coastal lowland lying between the Huntington Beach and Newport Mesas. This gap was formed by the Santa Ana River, which begins high in the San Bernardino Mountains and flows over 100 miles southwesterly to discharge into the Pacific Ocean at Huntington Beach.

The gap is an alluvial valley about 2.5 miles in width and extends about 4.5 miles inland. Its surface elevations range from sea level at the coast to about 25 feet at its inland portions, while the adjoining mesa surfaces have elevations ranging from 50 to 110 feet above sea level.

During the 1890s, agricultural interests were attracted to the flat fertile surface of the Santa Ana Gap, where artesian wells yielded water of excellent mineral quality. Until about 1920, water flowed freely from these wells. By the mid-1920s the increased production of ground water had led to the lowering of pressure levels in the shallow water-bearing zone to elevations below sea level. Consequently, encroachment of water from the ocean began to occur in the shallow zone, called the Talbert aquifer.

A wet period from 1936 to 1945 replenished the ground water basin and partially restored historic high water levels. During the period immediately following 1945, ground water was extracted in quantities that exceeded natural annual fresh water recharge, and a rapid decline in ground water levels ensued. In addition, upstream diversions from the Santa Ana River were reducing the flows to Orange County, resulting in less recharge to the basin. As the saline waters intruded from 1930 to 1960, a number of wells tapping the zones below the Talbert aquifer also began to experience intrusion.

The Orange County Water District

The Orange County Water District was formed in 1933 by a special act of the California legislature. The district has a broad authorization to protect and manage the ground water basin in Orange County. OCWD functions as a manager of the basin for those agencies that provide retail water service to consumers.

The district initially covered 163,000 acres inhabited by 60,000 people. Total water use in 1933 was 150,000 acre-feet, of which 86 percent was used for irrigation of agricultural land. Today the district covers nearly 220,000 acres and has a population of more than 2 million. Water usage has completely reversed since 1933, and urban use constitutes 94 percent of the district's total water demand. The basin supplies approximately 75 percent of northern Orange County's annual water demand, averaging 300,000 acre-feet. Although the basin contains between 10 million and 40 million acre-feet of water, its usable storage is limited by sea water intrusion and possible subsidence to approximately 1 million acre-feet.

Innovations to Prevent Seawater Intrusion

Recharging the Basin

With the importation of Colorado River water in 1940-1941 the district's water demands on the ground water basin were reduced. However, ground water levels continued to drop until 1954, when imported water was used to supplement the district's ground water replenishment program.

In 1956, with an accumulated overdraft of 705,000 acre-feet, water levels were at an historic low. The purchase of imported replenishment water escalated dramatically from approximately 80,000 acre-feet in 1957 to 235,000 acre-feet in 1963. More than 1.165 million acre-feet of imported water from the Colorado River was purchased for replenishment of the ground water basin during the period 1956 to 1965. After 1956, water levels began to recover and rose through 1964, despite the continuing drought.

The replenishment program was a success, reducing the accumulated overdraft to approximately 15,000 acre-feet. By 1964, average water levels in the basin were 24 feet above sea level, up from 20 feet below sea level in 1956 and equal to the average water level in the landmark year of 1944. Because of changes in the distribution of water in the aquifers, however, the average water levels inland were far above 1944 levels while those along the coast were far below what they were in 1944. Despite replenishment efforts, sea water intrusion continued along two areas of the coast, at the Alamitos Gap and the Talbert Gap.

Intrusion Barrier Projects

Together with the Los Angeles County Flood Control District, OCWD constructed the Alamitos Barrier Project located near the mouth of the San Gabriel River. By 1950 the ground water level in the Alamitos Gap, which straddles the boundary between Los Angeles and Orange Counties, was 30 feet below sea level. By the spring of 1962, sea water intrusion had proceeded more than 3 miles up Alamitos Gap. Barrier operation began in 1965 with 14 injection wells and has expanded to 26 wells.

An average of 5,000 acre-feet of imported water purchased from MWD is injected at the Alamitos Barrier Project each year. The barrier has halted salt water intrusion in the Central Basin located in Los Angeles County and the Orange County basin, protecting them from further degradation. Operation of the barrier continues to be a joint project of the Los Angeles County Department of Public Works and OCWD.

By the late 1960s, district officials recognized that sea water intrusion of the Talbert Gap could not be averted solely by replenishment of the basin through its recharge operations and began construction of the Talbert Barrier Project. To provide a supply source for the Talbert Barrier, an advanced wastewater treatment plant, Water Factory 21, was built in 1973. The project includes a 15 mgd advanced wastewater treatment plant and a hydraulic barrier system consisting of 23 multipoint injection wells with 81 injection points. At present, injection water for Water Factory 21 is a blend of 14 mgd reclaimed wastewater and 9 mgd of ground water pumped from a deep aquifer zone that is not subject to sea water intrusion.

Water Factory 21 treats secondary effluent using lime recalcination, multimedia filtration, carbon adsorption, disinfection, and reverse osmosis. All components of the reclamation system and hydraulic barrier facilities have functioned well since operations began in 1976. The quality of the injected water has consistently met or exceeded all health regulatory agency requirements.

With the completion of the two sea water intrusion injection barriers, the ground water levels in the two basins now can be safely kept below sea level, which allows for a more efficient ground water management plan. The Alamitos Barrier and the Talbert Barrier have effectively halted sea water intrusion in the basin so that it can be used as a ground water storage reservoir, providing more access to available local supplies.

The Value of Averting Sea Water Intrusion

The principal economic effects on an area where the ground water basin is subjected to seawater intrusion are the impairment of the basin as a storage reservoir, the degradation and loss of the potable water supply stored in the basin, and the loss of the basin's value as a fresh water distribution system. Each of these functions, which can be impaired or completely destroyed by sea water intrusion, has tremendous economic value in a large basin area such as Orange County. If protected from intrusion, this water supply would continue to be fully available for use.

The Value of the Basin as a Storage Reservoir

The absence of precipitation during summer months reinforces the seasonal variation in the demand for water in southern California. Furthermore, average annual precipitation is not only modest but also highly variable. Dry years often come in succession for a decade or more. Thus ground water basins have functioned as natural regulators of runoff and as storage reservoirs for daily, cyclical, and seasonal peaking requirements. These requirements must be met either from surface storage facilities or from ground water basins.

In southern California standby pump and well capacity is much more economical to develop and maintain than surface storage and distribution facilities. When the additional sizing costs necessary to meet peaking requirements in surface distribution facilities are considered, the critical economic importance of ground water basins for peaking purposes in southern California becomes apparent. If ground water storage is not continuously available for peaking purposes, alternative surface facilities would be required. Based on the present value and scarcity of land and construction costs, these facilities would represent a cost of hundreds of millions of dollars.

Underground storage is also preferable in several respects to storage in surface reservoirs. Water stored underground does not evaporate, as it does in surface storage and aqueducts. If the water needs to be stored for long periods, evaporation losses can be a serious concern, especially in arid regions where evaporation rates are high.

In addition, natural runoff that percolates into a ground water basin loses economic value if it flows into a basin degraded by sea water. This fresh water supply of approximately 270,000 acre-feet per year in Orange County would become unusable as a potable source.

Value of the Fresh Water Distribution System

The ground water basin acts as a distribution system because water may be extracted in a wide area overlying the basin. If the basin is lost, then a distribution system to deliver the alternative surface supply to the consumers must be constructed. In addition, the abandonment of the capital investment in wells and pumping facilities would represent a substantial economic loss.

Value of Potable Water Supply in Basin

The dependency on imported sources is becoming less desirable for southern California. The Metropolitan Water District provides the region with two sources of imported water. One is from northern California through the State Water Project and the other is from the Colorado River. Environmental concerns over the San Joaquin/Delta River system have had an impact on State Water Project

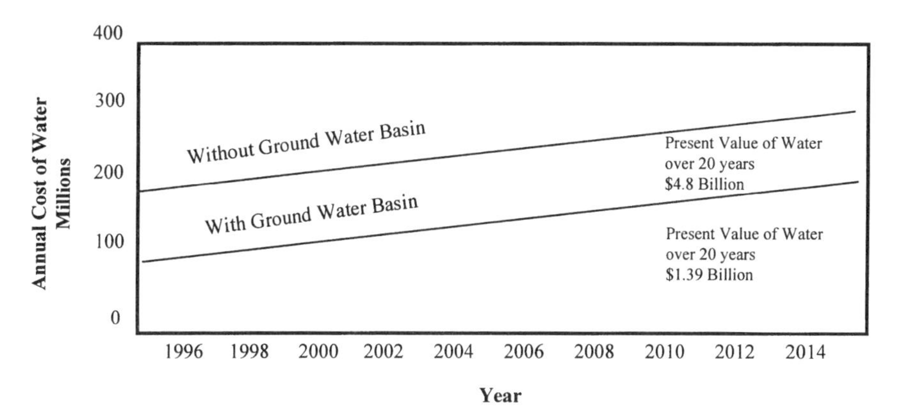

FIGURE 6.1 Estimated annual cost of water to retail producers: With or without ground water basin.

resources, and Arizona and Nevada are looking to increase their allotment of Colorado River water. Access to Orange County's valuable local ground water resource decreases the district's dependence on this more costly, less reliable imported water supply.

Ground water is generally less expensive than imported water, primarily because of the development and transmission costs of the imported supplies. As seen in Figure 6.1, it is projected that the value of Orange County's ground water over a 20-year period will be approximately $1.39 billion, and the value of imported water will be as high as $4.80 billion. Figure 6.1 shows the annual cost of water for the district with and without a ground water basin and indicates that the present value difference of the two scenarios is approximately $3.41 billion; this is one measure of the value of the ground water basin although it presumably represents a lower bound estimate of the true value.

Under current conditions with the ground water basin, retail producers within the district are able to meet approximately 75 percent of their demands by pumping from the ground water basin. The price of this water is estimated at $138 per acre-foot. which includes a pumping assessment of $85 per acre-foot and an energy cost of $53 per acre-foot.

In 1995 approximately 300,000 acre-foot of water was pumped from the ground water basin. Approximately 130,000 acre-foot of imported water was purchased from MWD. Of the water purchased from MWD, approximately 100,000 acre-foot was noninterruptible treated water at a price of $426 per acre-foot. The remaining 30,000 was purchased as seasonal shift water at a price of $286 per acre-foot. By having access to a ground water basin, retail producers are able to participate in the MWD seasonal shift program, which allows them to purchase imported water at a discount during the winter months.

The total cost of purchasing these three types of water (ground water, imported, and imported "seasonal shift") by the retail producers to serve their customers was approximately $92.6 million in 1995. Alternatively, if the ground water basin were not available, the entire 430,000 acre-foot of necessary supplies would have to be purchased from the MWD at the rate of $426 per acre-foot. The total cost of this water is approximately $183 million, which is roughly twice the water supply cost when the ground water basin is available.

In addition, the savings derived from the use of the ground water basin, compared to the cost of sea water intrusion facilities, offsets the cost of constructing and operating a barrier. For instance, capital and construction costs for Water Factory 21 were approximately $57 million (in 1995 dollars), with an average operation cost of $6 million per year.

Although a dollar value cannot readily be assigned to it, the value of the ground water basin for an emergency water supply and distribution system constitutes an important justification for protection. If the surface distribution system should become unusable because of a natural or human-made emergency or the imported supply were interrupted, reduced, or contaminated, ground water could eliminate tremendous economic loss or even assure survival. The value of an emergency supply would also be enormous during a period of extended drought. The value of the basin would increase with the severity and duration of the emergency.

Conclusion

Ground water and ground water basins are valuable resources because of the quantity of supply and the possibility for storage and distribution. William Blomquist stated in "Dividing the Waters," "When a ground water basin is destroyed, water users not only lose the comparative advantages of underground water storage and distribution, but they also suffer enormous financial costs. Replacement of ground water storage and distribution capacity, even if feasible, would be an economic disaster" (Blomquist, 1992).

Communities in Orange County would be forced to turn to more costly imported water to meet their water supply needs if the ground water basin were lost to sea water intrusion. To replace local ground water supplies in Orange County with enough imported water from the MWD of southern California to maintain current levels of use for a 20-year period would cost water users at least $4.8 billion dollars.

Reference

Blomquist, W. A. 1992. Dividing the Waters: Governing Ground Water in Southern California. San Francisco: ICS Press.

INCORPORATING THE VALUE OF GROUND WATER IN SUPERFUND DECISION-MAKING
Woburn, Massachusetts

Background

Knowing the value of ground water is important in evaluating remediation alternatives for Superfund sites involving ground water resources. Unlike other case studies where the primary concerns are maintaining the quantity and/or quality of the ground water, in the Superfund setting the ground water resource has been contaminated and the issue is primarily one of restoring quality. Thus the value of the ground water will be reflected in the values associated with moving from a situation where the aquifer is not usable to situations where some uses can be made of the aquifer.

In this section a brief overview is provided of a study on the costs and benefits of ground water remediation for a Superfund site located in Woburn, Massachusetts (Spofford et al., 1989). This study, conducted by Resources for the Future (RFF), illustrates the importance of valuing ground water as a component in a cost-benefit analysis of ground water remediation decisions and the complexities that such an assessment involves.

In 1980 the Comprehensive Environmental Response, Compensation, and Liability Act (CERCLA) was enacted to facilitate cleanup of the nation's worst hazardous waste sites. CERCLA and the subsequent 1986 Superfund Amendments and Reauthorization Act (SARA) created a fund, the Superfund, to pay for site cleanup when parties who caused the contamination could not be found or could not pay for the cleanup themselves.

Many of the Superfund sites involve contamination of ground water resources or are listed as potential threats to local public water supply wells (Canter and Sabatini, 1994). Contaminants found in public ground water supplies are mainly volatile organic contaminants (VOCs) such as TCE (trichloroethylene), PCE (tetrachloroethylene), 1,2-DCE (dichloroethylene), vinyl chloride, and benzene; other contaminants commonly present in various combinations included heavy metals (chromium, lead, and arsenic) and polynuclear aromatic hydrocarbons (PAHs).

Economic considerations should play a key role in determining the extent to which ground water remediation is pursued at these sites. If the expected costs of cleanup exceed the expected benefits of improving the quality of the ground water, then the economically rational alternative is not to remediate the ground water. This might be the case if the contaminated ground water resource is not used and is perceived to have a low potential for future use or if there are relatively low-cost substitutes. Alternatively, there may be different levels of remediation depending upon the future uses and expected costs of the ground water resources. Thus, in the context of Superfund remediation, the value of ground water is reflected in the benefits associated with improvements in the quality of the contaminated aquifer, and the value will depend upon the level of remediation. In addition, the benefits of remediation cannot exceed the total economic value of the ground water for that level of remediation.

Analysis of the Woburn Superfund Site

The Superfund site in east Woburn, Massachusetts, included two municipal water supply wells, which had been found to be contaminated with chlorinated solvents in 1979. Prior to contamination, the local aquifer segment had served as a drinking water supply for the city of Woburn and as a water source for several industrial users. Possible levels of remediation included restoring the aquifer to drinking water quality as well as maintaining the aquifer for nonconsumptive purposes only.

Because of measured contaminant concentrations and related health concerns, TCE had been identified as the key contaminant for remediation planning. The costs of remediation that were analyzed in the RFF study were based on withdrawal of the contaminated ground water from the aquifer using extraction wells, treatment using aeration towers, and return of the treated water to the aquifer using injection wells.

In estimating the benefits of remediation, the RFF study considered only those benefits that were associated with direct use of the aquifer. The benefits that were not measured in the RFF study but that should be considered in a study designed to measure the TEV of the contaminated aquifer include reductions in losses of recreational opportunities; reductions in ecological damages; reductions in losses of intrinsic value, including bequest and existence values; reductions in health damages due to morbidity as opposed to losses due to mortality (these were included in the assessment of health damages in the RFF study); and reductions in fear and anxiety associated with switching from a water supply that is perceived to be safe (bottled water or a municipal water source) to a water supply that may be perceived as unsafe (the remediated ground water). As the authors indicate, estimates of these benefits were not included in the report since the primary purpose of the research was to illustrate the impact of uncertainties on measures of net benefits as opposed to replicating the true benefits and costs for a specific site, and study funds were limited (Spofford et al., 1989).

To estimate the benefits of remediation, the researchers hypothesized two management contexts. The first is fairly simple and basically involved using an alternative water supply for the entire city of Woburn. The contaminated ground water supply was assumed to be replaced with water purchased from the Massachusetts Water Resources Authority (MWRA), which was the least expensive alternative available to the city. The additional cost per gallon of using MWRA water as opposed to pumping the aquifer multiplied by the total use of water in the city provides a lower bound on the value of the aquifer in a given year. (The study estimated this cost to be approximately $0.32 per 1,000 gallons in 1986.) This method of valuing the ground water resource is basically a replacement cost approach, which does not reflect the values people may attach to clean ground water or the disutility attached to knowing that the aquifer is contaminated.

The second management context developed in the study was more complex and was based on the underlying assumption that all the households that had previously relied on private wells as opposed to a municipal water supply system would continue to use "contaminated" water, although the extent of uses by these households could vary. The study hypothesized a range of situations from ones in which consumers avoided using the water for drinking, food preparation, and personal hygiene, in which case the direct use value of the ground water would be reflected in the costs of purchasing alternative drinking water supplies, to situations where individuals continued to consume the contaminated ground water. In the latter situation, the direct use value of the ground water would be reflected in the health costs, or damages, associated with the consumption of contaminated water.

To implement this second management context, a model of consumer decision-making under uncertainty that incorporated perceived health risks associated with consumption of contaminated water and the costs of alternative drinking supplies (such as bottled water) serves as the basis for determining how individuals made their consumption choices and for constructing a demand curve for ground water. The value of ground water can then be measured as the area under the demand curve. The consumers' demand for ground water depends upon many factors, including the normal demand determinants (income, price of substitutes, etc.) as well as attitudes toward perceived health risks and the levels of TCE in the water. Over time, the demand for ground water may shift as these underlying factors change, and thus the value of the ground water will change.

The direct use value of the ground water in this second management context will depend upon the extent to which the contaminated ground water is consumed as a source of drinking water. For example, if the water is not consumed as a source of drinking water, then the direct use value of the ground water would be reflected primarily in the costs of bottled water, which substitutes for the human consumption component of per capita water consumption. RFF estimated this level of use to be approximately 3.65 gallons per person per day out of a total use level of 130 gallons per person per day. In this context the direct use value of ground water can be estimated using the estimates of avoidance costs, but as in the first management context, this value does not reflect any of the indirect or nonuse values of the aquifer.

If consumers continue to drink the contaminated water from the private wells, there are no avoidance costs per se, and the value of the ground water is reflected in the health costs associated with its consumption as drinking water plus the indirect or option value. These two extremes provide some bounds on the direct use value of ground water or, alternatively, provide bounds on the average benefits of ground water remediation for the management context where households in the area still continue to use private wells.

As the RFF study noted, estimates of the benefits of ground water remediation depend on many assumptions. For example, in specifying the model of consumer behavior, assumptions are needed regarding the level of the potential health damages, discount rates, concentrations of TCE over time in the aquifer, and future costs of alternative drinking water supplies. These assumptions affect decisions on avoidance costs and measures of health damages and thus the direct use value attached to ground water by individuals who have private water supply wells.

An additional source of uncertainty relates to the behavior of individuals confronted with different levels of TCE in drinking water that exceed the standard. The extent to which individuals will avoid contaminated well water and their willingness to pay for such avoidance varies with perceived risks to health of different levels of TCE in the drinking water. The authors cite the lack of an adequate methodology for measuring perceived health risks as a major limitation in using this approach to quantify the value of ground water or using this management context as a basis for remediation decisions (Spofford et al., 1989).

The general findings of the RFF report indicate that for the first management context the net benefits of remediation were positive, indicating that from an economic perspective it is more efficient to remediate the aquifer than to continue using an alternative water supply. This finding also held true for the second management context, where it was assumed that all the households affected by the contaminated aquifer had previously relied on private wells: it is more efficient to remediate the ground water than it is to permit the exposed population to continue to use contaminated well water. Comparisons among the net benefits for alternative remediation designs would shed some light on the relative efficiency of alternative cleanup options.

Conclusions

Several conclusions pertaining to the value of ground water can be drawn from the Woburn case study:

- Economic valuation of ground water for the specific hazardous waste site is crucial to making informed decisions regarding the status of remedial action. The conclusions reached will be site specific, depending on the nature of the contaminant and the current uses of the aquifer.
- Determining the full economic value of the aquifer will often be difficult because of the indirect nature of many of the benefits. However, assessing the direct use benefits poses a much simpler task and may serve as a lower bound on the benefit estimates.
- Technical and economic uncertainties must be recognized in quantifying

the value of the ground water resource. While the RFF study noted many uncertainties, those that pertain to the benefit side of the equation are substantial enough to warrant further research.

References

Canter, L. W., and D. A. Sabatini. 1994. Contamination of public ground water supplies by Superfund sites. International Journal of Environmental Studies, Part B 46:35-57.

Spofford, W. O., A. J. Krupnick, and E. F. Wood. 1989. Uncertainties in estimates of the costs and benefits of ground water remediation: Results of a cost-benefit analysis. Discussion Paper QE 89-15. Washington, D.C.: Resources for the Future.

APPLYING GROUND WATER VALUATION TECHNIQUES
Tucson, Arizona

Background

The objective of this case study is to illustrate how incorporating the economic concepts and techniques developed in Chapters 1 through 4 of this report can assist in management of ground water resources over the long term. Unlike the previous case studies, which are limited to reviews of existing work and demonstrations of value of ground water in various contexts, the Tucson case study illustrates the application of the conceptual valuation framework described in Chapter 3.

The Tucson case study is notable both for the diversity of ground water services it illustrates as well as for the urgency of policy attention the area's water management system requires. The intent is not to calculate the incremental change in value of services provided by ground water in the "with treatment" and "without treatment" condition but to identify the steps required to implement the valuation process in a real-world context. This case study simplifies and abstracts information from the actual Tucson situation in order to better illustrate the role of economic valuation in improving management of ground water resources.

Ground water provides numerous extractive services in Tucson, including residential, commercial, agricultural, and industrial water uses. The region's ground water resources also provide a range of *in situ* services, such as prevention of land subsidence, reservoir functions that will buffer future drought associated with shortages in surface water supplies, bequest value, and ecological services such as maintenance of riparian habitat. Policy-makers in Arizona have struggled to reduce the extent to which ground water supplies in the region are mined. Ground water policies have been put in place as a mechanism to ration and conserve supplies for future use. Alternative renewable surface water supplies to augment and substitute for ground water have been developed at great cost in anticipation of future demands.

As is the common practice, the price of ground water in Tucson does not reflect any of the commodity values, including the extractive and *in situ* service flow values. It is based on the cost of distribution, including capital, operations and maintenance, and administrative costs. Ground water is thus the least expensive and highest-quality water supply available. In a dynamic pricing environment, water would be priced to incorporate marginal extraction cost and user cost and would reflect the values of all use and nonuse service flows.

Instead of relying on price to ration scarce ground water supplies, Arizona water managers have focused on regulations and other nonprice policies to reduce water use. The total economic value of ground water supplies in any location is affected by the institutional, policy, and hydrological constraints that shape current and future use and define management options. Policy-makers must recognize this institutional and political context in order to make an accurate assessment of the services ground water provides.

Tucson's Water Resources

Tucson has relied on a high-quality ground water supply to meet all of its demands for water. Ground water use has exceeded natural recharge (precipitation and return flows) annually since 1940, leading to a situation in which over half of annual use is from mined ground water. In Tucson's desert climate, there are no viable local renewable surface supplies (other than municipal effluent) to substitute for ground water resources.

Although there is a substantial amount of ground water in the aquifer, dependence on mined ground water has a number of negative consequences. Falling ground water levels have eliminated many of the free-flowing rivers, streams, and associated riparian habitat in most of southern Arizona. The risk of subsidence with continued depletion of ground water is quite severe in the central Tucson wellfield that underlies the city of Tucson; a worst-case estimate is that the ground level will drop by 12 feet by 2024 (Hanson and Benedict, 1994). In addition, the most productive parts of the aquifer are nearly exhausted, which can be expected to lead to substantial increases in pumping costs. As a consequence of municipal pumping in the central Tucson wellfield, ground water levels have fallen as much as 170 feet.

Legal and Institutional Constraints

Legal and institutional constraints on ground water use frame the valuation context. The Tucson Active Management Area (AMA) is one of five AMAs in the state established pursuant to the 1980 Groundwater Management Code. The Tucson AMA has a statutory goal of safe yield by 2025. The safe yield goal requires that the amount of ground water used on an average annual basis must not exceed the amount that is naturally or artificially recharged.

The code established stringent limitations on ground water use within AMAs. Farmers receive an allocation based on historic cropping patterns assuming maximum irrigation efficiency. No irrigation of new agricultural land is allowed. Allocations to municipal water providers are on the basis of their average historical use in gallons per capita per day. A "reasonable" reduction is required within each water company, based on an evaluation of conservation potential. All large industries are directly regulated, using an approach based on either allotment (for golf courses) or best management practices (for copper mines, sand and gravel, electric power, etc.).

In addition to demand management policies, there are economic incentives to discourage development of new ground water uses and encourage use of renewable supplies, primarily imported surface supplies (from the Central Arizona Project, or CAP) and wastewater effluent. One of the primary tools for moving from the current state of overdraft to the safe-yield condition is the 100-Year Assured Water Supply (AWS) Program. This program, administered by the Arizona Department of Water Resources, severely limits the amount of ground water that can be used for municipal purposes. The cumulative amount of ground water that the city of Tucson can legally withdraw as part of its AWS is approximately 3.5 million acre-feet. If the city were to rely solely on ground water for its supply, its cumulative ground water withdrawals would exceed this amount before 2030. Without utilization of Tucson's CAP allocation, Tucson would not qualify for a designation of AWS.

Demand for Water

The population of the Tucson AMA is estimated at 750,000 for 1995 and is projected to reach 1.3 million by 2025. The majority (78 percent) of the population in the Tucson AMA is served by Tucson Water, the water utility operated by the city of Tucson.

Total water use in the AMA is currently close to 300,000 acre-feet per year (see Table 6.2); more than half of the total water supply is mined ground water.

TABLE 6.2 Tucson AMA Water Demand

Sector	1994 USE (in acre-feet)	Percent
Agricultural	97,900	31
Municipal	148,500	47
Industrial	18,600	6
Mining	45,000	14

SOURCE: Arizona Department of Water Resources, 1996.

Under current population projections, total demand for water in the Tucson AMA is expected to be approximately 427,000 acre-feet per year by 2025 (Arizona Department of Water Resources, 1995 and 1996), increasing by 50 percent from the 1995 levels.

Development of Alternative Renewable Water Supplies: The Central Arizona Project

The Central Arizona Project is a 330-mile canal built by the U.S. Bureau of Reclamation, with pumping stations and associated distribution and flood control facilities. It extends from Lake Havasu to Tucson; the total cost, including federal, local, and private investment, exceeds $4 billion. A major feature that has not been constructed is a reliability feature for Tucson, a terminal storage reservoir. Tucson has the largest municipal allocation of Colorado River water—148,200 acre-feet.

Applying the Economic Valuation Framework in Tucson

Policy Constraints on Use of Renewable Supplies

At the end of 1992, nearly half of Tucson Water's customers (84,000 metered connections) began receiving CAP water. After unanticipated water-quality problems arose (rusty water, turbidity, taste and odor problems, and bursting pipes), 37,000 metered connections in the older parts of town were returned to ground water in October 1993. Water-quality problems were attributed to old cast iron and galvanized steel water mains and household plumbing, combined with pH and other chemical attributes of the imported surface water that encouraged corrosion.

In January 1995, the Tucson City Council voted not to directly deliver CAP water to customers until the water-quality problems were fixed. On November 7, 1995, the citizens of Tucson approved a citizen's initiative (Proposition 200; the Water Consumer Protection Act) prohibiting direct delivery of CAP water to customers unless it is treated to the same quality as ground water for hardness, salinity, and dissolved organic material. This can be accomplished only through advanced treatment, probably reverse osmosis or nanofiltration.

Defining the Management Options Within Current Constraints

Under the existing constraints, Tucson's water planners must define feasible options for meeting Tucson's present and future water needs. These options are constrained by the above policies and laws.

In this application of the conceptual framework, recharging the untreated CAP water supplies into overdrafted aquifers is the base or "without-treatment" case. Water to meet all demands would continue to be pumped from ground water supplies in a conjunctive management scheme. Treating surface water with advanced membrane filtration to remove salinity and organic material prior to direct delivery to customers is the "with-treatment" option. It is important to note that in this example the valuation techniques are not used to calculate the TEV for ground water. Rather, they are used to evaluate the change in ground water value that results from a particular policy decision.

Identifying Changes in the Quantity and Quality of Ground Water

Initially, hydrologists must establish the quantity and quality of Tucson's ground water resources. Policy-makers need to assess how the "with-treatment" management option would change this baseline quantity and quality. Since the recharge option is considered the baseline, an accurate assessment of the impacts of artificial recharge potential is also needed.

The quality of the water that is pumped depends on where the CAP water is recharged relative to the location of recovery, the nature of the aquifer materials, the degree of blending with local ground water, the distance the water travels in the subsurface, and the presence of any source of contamination. The Groundwater Management Code allows an entity to recharge in one location and recover at a distant location within the same AMA, provided certain criteria are met.

Identifying Changes in Service Flows

The next step is to link the management decision with the changes that result in the time path of services that the ground water will provide under the alternative. This is where the critical input from scientists and hydrologists is required. The "without" scenario describes the services provided in the base case and the incremental changes that result from substituting treated surface water supplies.

Incremental Changes in Extractive Service Flows

Although Colorado River water is viewed as a high-quality water source for millions of people in the Southwest, there are several ways in which recharge using CAP water can reduce the quality of the water available for extractive uses. CAP water as treated with conventional surface water treatment methods meets all of the EPA maximum contaminant levels (MCLs), but the aesthetics, taste, and hardness of CAP water were a major issue for Tucson Water when the supply was directly delivered to customers from 1992 to 1994. Any use of CAP water in the basin, whether through direct delivery or recharge, will increase the salinity level of the aquifers within the Tucson AMA. The only way to avoid the increase in salinity is to utilize an advanced treatment approach (probably using membrane technology) to remove the salts. This technology is expensive; it is therefore important to identify the value of the changes in service flows that would be provided to decide whether the additional expense is justified. Unlike native ground water, surface water tends to contain pathogens, some of which are difficult to remove.

CAP water has roughly twice the total dissolved solids (TDS) and salinity of the local ground water, and it contains organic precursors that can, in combination with chlorine, cause formation of trihalomethanes, a group of chemicals known as carcinogens. Depending on the contact time and travel through aquifer materials, the recharge process may reduce the organics and disease-causing organisms (bacteria and viruses), but it does not affect the salinity and hardness of the water. Therefore, recharge of untreated CAP water is likely to influence the quality of the water in the aquifer. To the degree that this same water is recovered for delivery to municipal customers, costs for end users will increase, because higher salinity and hardness translate into the need to replace appliances more frequently and increase the maintenance of irrigation and cooling systems.

Depending on the location of the recharge, impact on water quality may not be substantial. For example, there are areas in the AMA where the end users may not experience negative effects from the higher salinity (agriculture usually has few problems with 700 mg/l TDS). However, it is important to note that the salt load brought in with the CAP water will be distributed in the vicinity of the recharge facilities and could migrate over time to surrounding aquifer materials unless the withdrawal facilities are in the same location.

Recharge will have a positive effect on extractive values if the water is recharged in the vicinity of wells supporting extractive uses. However, several of the prime recharge locations are not near the central Tucson wellfield.

The treatment option requires the development of an advanced water treatment facility, probably nanofiltration or reverse osmosis, to remove the salts, organics and solids as required by Proposition 200. Aside from the capital cost of the facility, which is several hundred million dollars, a major concern is disposal of the brine stream. Depending on how this salt-laden wastewater is directed, the effect on ground water service flows varies. The brine stream from such plants is normally discharged into surface water or injected into deep wells. Neither of these options is available in Tucson. The most likely disposal alternative is evaporation ponds, with the sludge deposited in lined landfills.

The advanced treatment option provides the highest-quality water for municipal uses. It would not affect the quality or quantity of water for agriculture or mining. Direct delivery has many benefits, since it leaves the ground water in place and should allow for at least partial recovery of all of Tucson's wellfields. Advanced treatment will limit the avoidance costs of many municipal end users, who would otherwise buy bottled water, resort to point-of-use treatment devices, or replace their appliances more frequently as a result of using CAP water either directly or after recharge. Membrane treatment will also eliminate the possibility of *Cryptosporidium* or *Giardia* outbreaks, if the treated water is blended with ground water rather than surface water before delivery to customers. Blending of membrane-treated water is generally recommended to improve the taste, reduce corrosiveness, and reduce costs.

Depending on the brine stream disposal method (most likely through evaporation ponds), there could be localized impacts on water quality in the aquifer. Another option is to deliver the brine to existing wastewater treatment plants, to be blended with less salty effluent before discharge. This would result in high salinity downstream from the wastewater facilities. The ground water quality in these areas could be degraded, affecting service flows.

Incremental Changes in In Situ *Service Flows*

In situ service flows can be categorized as use and nonuse. For Tucson, *in situ* uses include use of the stock to: (1) assimilate contaminated runoff from extractive uses and attenuate existing areas of ground water contamination; (2) buffer future drought on the Colorado River system in a conjunctive use scheme; and (3) support the soil structure in the aquifer and prevent subsidence. Additional *in situ* services include: existence value (based on a desire to protect the aquifer as part of the natural system); bequest value (the intent to protect water for future generations); and ecological services in which ground water supports surface water flows and riparian habitat.

Recharge Effects

If recharge is used to limit water-level declines in areas that are prone to compaction, it will help support *in situ* uses. There are two ways to limit the subsidence potential in the central Tucson wellfield: reduce the amount of future pumping there by withdrawing ground water elsewhere and recharging within the central wellfield. The former is easier in this case, since Proposition 200 precludes an effective way to recharge in the central basin (injection recharge). Surface recharge in areas of subsidence can actually accelerate subsidence, since its weight adds stress to the aquifer materials.

Recharge results in the storage of water for future use, which increases the buffer value of the aquifer. If the water is available for future generations, then it supports the bequest value as well. Those who stress the existence value of the aquifer would likely prefer that higher-quality ground water be maintained rather than degraded by CAP water through recharge. However, it is not clear what the quantity/quality trade-off is for existence value.

If recharge occurs in the vicinity of streambeds, it is likely to support riparian habitat or provide for an expansion of habitat values. Recharge facilities can easily be designed with a habitat/recreation component, guaranteeing a positive impact on ecological values. However, there are costs associated with increasing habitat values, particularly if threatened or endangered species become a component of the new habitat. The costs are associated with endangered species regulations, which could require permanent maintenance of the artificially created habitat to protect a particular species. This introduces a cost associated with irreversibility—the decision to recharge could be legally required to continue even if another water use option were more desirable from other perspectives.

Advanced Treatment Effects

Substitution of treated surface supplies for pumped ground water means that most of the wellfields in the vicinity of key riparian areas would not be used often. In addition, the regional water tables should rise in the wellfields because of natural recharge. Both of these occurrences should increase the quantity of water available for ecological service flows.

The direct delivery option with advanced treatment protects both the quality (depending on the disposal of the brine stream) and the quantity of water in the aquifer. The buffer value of ground water would be the highest in this option, since there will definitely be future supply shortages, during which consumers will rely on ground water. By ending the current pumping in the central wellfield, additional subsidence is likely to be avoided. Ground water will be available for future generations, and those who value existence of the aquifer would have the quality as well as the quantity protected (at least in theory).

Valuing Changes in Extractive Service Flows

Incremental Extraction Costs (Marginal Benefits of Quality Changes). In the Tucson example, the difference in the extractive service flows between the two options is related primarily to the reduced pumping costs and the reduced number of wells required to serve the community, as well as the water-quality issues caused by recharge of untreated CAP water that occurs in the one option. If the recharge does not occur in the vicinity of existing wellfields, water levels will continue to decline in those areas. Lowering the water level has two economic effects: it increases the amount of energy required to pump each acre-foot of water, and it results in the need to drill more wells since the most productive part of the aquifer may be exhausted. Direct delivery after treatment eliminates these costs because there would be little dependence on ground water as a supply except during infrequent CAP shortages and canal shutdowns.

The only methods identified for evaluating the change in services related to pumping costs for extractive purposes were standard engineering analysis techniques—increased energy costs associated with increased head and well drilling and system extension costs. The major issues associated with this type of analysis are uncertainty and discounting. It is unclear how productive deeper parts of the aquifer will be and how many wells will be required to replace the capacity of the existing high-capacity wells. The time element is also uncertain; it is not known how many years will pass before expanded infrastructure is required to maintain current production levels. Since the growth rate on the city's system is 1-2 percent per year and long-term outages may occur on the CAP canal, pumping capacity must be expanded to meet the increased demand as well.

Incremental Quality Costs (Marginal Benefits of Quality Changes). There is an additional reduction in service flows as the water level drops (assuming that recharge does not occur in the vicinity of production wells). The water that is withdrawn at greater depth in the Tucson basin is higher in salinity and TDS. This lower quality, like that of CAP water, will increase costs for end users in the residential, commercial, and industrial sectors. Households and firms may buy bottled water or in-home treatment devices (avoidance costs) or replace appliances more frequently. Industries and individuals with private wells will be similarly affected. These costs do not occur in the case of the direct delivery with advanced treatment option.

Possible techniques to evaluate the costs associated with reduced water quality include the averting behavior method and the contingent valuation method. To the extent that wellhead treatment or well replacement is required, standard engineering techniques must be used.

Opportunity Costs (Marginal User Costs). Ground water in the Tucson area is essentially a stock resource, since it is not naturally replenished at a high rate. Using a unit of ground water today means that it will not be available for future use. Methods that can be used to measure this "dynamic" opportunity cost include dynamic programming and intertemporal (optimal) control techniques. Although these methods are limited to measuring use values, they may be valuable in evaluating alternative options.

Valuing Changes in In Situ *Service Flows*

Subsidence Avoidance. The risk of subsidence is high in the central Tucson basin, where 60 percent of the city's water supply is currently pumped. In the recharge option, some pumping in the central wellfield would continue without replenishment in the same location. Subsidence costs include disruption of all utilities (sewer, water, electric, gas, etc.); damage to roads and buildings; and a possible permanent reduction in storage capacity of the aquifer. The direct delivery with treatment option eliminates the pumpage in the central wellfield, thereby essentially eliminating the possibility of increasing the rate of subsidence.

There are two techniques that can be used to measure the benefits associated with subsidence reduction. To the extent that utilities must be repaired or rerouted and roads and buildings must be repaired, standard production cost estimates can be prepared. The other method is the hedonic price (property value) method, since some areas are at considerable risk of subsidence and others are unlikely to experience any damage. Differences in market prices across these zones may begin to reflect these costs.

The degree of uncertainty associated with predicting subsidence damage is high. It is unclear how long it will take after an aquifer is dewatered for the compaction to occur. It is also unclear whether the whole basin will settle as a unit or whether it will settle differentially, causing subsidence cracks and substantially more damage. The normal pattern is that the cracking occurs near the edge of the basin, and downtown Tucson is near the base of the Tucson mountains. A risk/probability assessment may be required.

Reservoir Function. In some aquifers subsidence causes irreversible damage to the water-holding capacity because rewetting these areas fails to have any rebound effect. The irreversible aspects of subsidence need to be taken into account, at least from a qualitative perspective. Engineering analyses can be used to compare lost storage capacity to the costs of alternative storage facilities, such as reservoirs.

Buffer Value. A ground water value that is lost under the recharge option is the ability to buffer the effect of drought on the CAP system. This value is not as high in the recharge option, since all customers are receiving pumped ground water and continuous delivery is not critical. If a direct delivery option were selected in the future, however, the buffer value of the aquifer would have been lost if most of the ground water supply in the vicinity of the wells had been removed.

Methods associated with estimating buffer value, such as intertemporal optimization, may be applied (Tsur and Graham-Tomasi, 1991).

Existence Value. Certain values associated with maintaining the ground water aquifer intact are unrelated to any function or service the aquifer provides. This value is difficult to describe, so methods of estimating it are limited. The most likely technique to establish this value is the contingent valuation method.

Habitat Values Related to Water Quantity. Higher water levels in the vicinity of some proposed recharge sites are likely to enhance habitat values. In addition, recharge sites can be designed with habitat enhancement as a component. However, in comparing the two options, it should be noted that it may be possible to create habitat using the brine stream from the advanced treatment facility. The options for improving local habitat due to recharge are offset by the probability that existing mature riparian habitat (such as the Tanque Verde Creek area) in the Tucson basin could be destroyed by continued pumping of the ground water.

Because impacts on habitat are visible only in limited areas, the hedonic price method (based on differences in property values) may be useful. Other methods that can be employed for evaluating the recreational/aesthetic values of habitat include contingent valuation and travel cost. Sources of uncertainty include lack of definitive information about the relationship of ground water level and habitat quality, the length of time it will take for dewatering to occur, and the limited number of remaining high-quality habitats to evaluate.

Habitat Values Related to Water Quality. The recharge of untreated CAP water will increase the aquifer salinity in the vicinity of the recharge site and in areas that are down-gradient from the recharge site. It is unclear whether the increased salt levels will have any negative effect on the development or maintenance of high-quality ecosystems. However, it is unlikely that salinity and TDS concentrations in the CAP will have a measurable effect on mature vegetation. If it can be demonstrated that mature riparian vegetation or mammals and birds are affected, it is possible that salt-avoidance ecological values can be measured using the contingent valuation or travel cost method.

Conclusions

Based on this descriptive approach for applying the valuation framework presented in Chapter 3, the following conclusions can be drawn:

- The treatment option is likely to have a higher benefit/cost ratio when the TEV of ground water is considered.
- Engineering analyses (changes in production costs) may continue to be used to establish costs where extractive services are a large component of cost, so long as costs are assessed at both the household level and the utility level.
- Quality issues may represent a more crucial impact on service flows than do quantity issues.
- There are any number of scientific and economic uncertainties associated with the use of ground water valuation methods.

References

Arizona Department of Water Resources. 1995. Proposal to Increase the Use of Colorado River Water in the State of Arizona. Arizona Department of Water Resources, Tucson, Arizona.

Arizona Department of Water Resources. 1996. State of the AMA: Tucson Active Management Area. Arizona Department of Water Resources, Tucson, Arizona.

Hanson, R. T., and J. F. Benedict. 1994. Simulation of ground water flow and potential land subsidence, Upper Santa Cruz Basin, Arizona. U.S. Geological Survey Water Resources Investigations Report 93-4196.

Tsur, Y., and T. Graham-Tomasi. 1991. The buffer value of ground water with stochastic surface water supplies. Journal of Environmental Economics and Management (21):201-224.

LESSONS LEARNED

Even though the case studies presented in this chapter are diverse, they share themes that provide the basis for observations and lessons.

- Each study involves unique hydrogeological features, ground water quality, uses of the resource, institutional requirements and constraints, and political contexts. Although the principles of the valuation framework described in Chapter 3 can be transferred, then, limited opportunities exist for transfer of benefits in subsequent studies.
- The case studies clearly demonstrate a range of ground water services even if comprehensive valuation studies have not yet been accomplished. Trade-offs in decision-making can be made based on descriptive (qualitative) information about this range. Clearly, quantification of the values of such services would provide more complete information for decision-making.
- These case studies highlighted the extractive value (service) of ground water. Several studies, however, recognized other services and TEV, and some focused on changes in value at the margin. Accordingly, ground water valuation studies should consider all components of TEV even though not all can currently be quantified. This approach would provide more complete information for subsequent decisions.
- Several cases illustrate the classic natural resource scarcity phenomenon where market indicators have been given only limited consideration relative to depletion. A price regime that more nearly mimics the market appears to be at least one of the ingredients of more rational water management. If that course is to prove fruitful for the long run, we probably must find a way to unbundle water demand and supply among such major extractive uses as drinking, bathing, laundry, lawn watering, and car washing—the variety of water uses likely to be valued very differently. Unbundling could involve dual water systems, or a single system with special treatment measures for drinking water, or creative water use accounting schemes.
- Some of the case studies identified concerns associated with human health risks from extractive uses of contaminated ground water. These concerns underscore the importance of carefully designed epidemiological studies, though even these are scarcely conclusive by themselves. Further, in the absence of epidemiological studies or information, debate will continue regarding actual and perceived health risks associated with degraded ground water.
- Ecological services provided by ground water are recognized in several cases; however, there appears to be a dearth of information on how to quantify and value ecological benefits. This need can be further emphasized by considering the contributions of ground water to the base flow of streams, maintenance of wetlands and their associated hydrological and biological functions, and the provision of riparian habitat.

• Technical and economic uncertainties must be recognized in efforts to develop site-specific ground water valuation information. Each of the case studies provides illustrations of such uncertainties. For example, in ground water valuation studies for Superfund sites, stochastic modeling of the contamination problem and potential effectiveness of cleanup measures should be used to develop ranges of resultant information that can be viewed as a type of "sensitivity analysis." Decision-makers should also consider the possible influences of uncertainties and nondelineated costs and benefits (of ground water services) as they interpret information.

• Ground water valuation studies must recognize the importance and limitations of the institutional and political context, which can lead to conflicts in conducting valuation studies and interpreting their results and influence on subsequent policies and decisions.

• Planning and implementation of valuation studies require the interdisciplinary efforts of hydrogeologists, engineers, environmental scientists, and economists, who must be able to interact with and learn from related disciplines.

APPENDIXES

APPENDIX A

Glossary

If you wish to converse with me, define your terms.
Voltaire

aquifer—an underground geologic unit that stores ground water.

aquifer capacity—the amount of water that is stored in an aquifer; sometimes used loosely to indicate the amount of water an aquifer can deliver under a specified set of pumping conditions.

averting behavior model (ABM)—an assessment approach in which costs incurred by households to offset or mitigate environmental hazards (e.g., expenditures on water purifiers to remove pollution from ground water) are used to infer value of clean water (Young, 1996).

base flow—that portion of a stream's flow derived from ground water (as opposed to surface runoff and interflow).

basins and watersheds—areas of drainage in which all collected water ultimately drains through a single exit point. Basins differ from watersheds only in the perception of their size: basins are usually considered to be much larger, composed of many watersheds. Within a watershed or basin, water moves both on and below the surface. Topographic "highs" prevent surface water from crossing from one watershed (aquifer) to another.

benefits—the gains, often measured as the sum of the monetary values of the direct and indirect uses, associated with the use of a resource or with improvements in the quality of a resource unit.

benefit-cost analysis (BCA)—a technique to compare the economic efficiency of different alternatives, usually applied to individual projects or policies. A BCA comprises the gross benefits of a project or policy and with its the opportunity costs. Benefits and costs are measured as changes in consumer and producer surpluses accruing to individuals in society.

bequest value (BV)—a willingness to pay to preserve the environment for the benefit of one's descendants.

buffer value—the difference between the maximum value of a stock of ground water under uncertainty and its maximum value under certainty where the supply of surface water is stabilized at its mean. Thus, this value arises from the ability of ground water to provide supplemental water supplies during short-term periods of drought or other supply disruptions.

cone of depression—a localized depression of the water level in an aquifer as a result of pumping. The depression may be confined to a small area or spread over a large area depending on the pumping rate and transmissivity of the aquifer. See Figure 2.4.

confined aquifer—an aquifer bounded above and below by units of distinctly lower hydraulic conductivity and in which the pore water pressure is greater than atmospheric pressure. An **unconfined** aquifer is not bounded above and is the uppermost aquifer. See Figure 2.2.

consumptive use—a use of water that does not return water (polluted or not) to system. For example, drinking water is consumed, while shower water returns to the treatment system.

contingent valuation method (CVM)—a method determining money measures of change in welfare by describing a hypothetical situation to respondents and eliciting how much they would be willing to pay either to obtain or to avoid a situation. Although well accepted for use values, CVM has many limitations when used to calculate nonuse values (Young, 1996).

costs—the mirror image of benefits, that is, what is lost when a resource increases or decreases in quantity or quality. In this context one can think of costs as damages. Costs also refer to the value of any resources used to change the quality or quantity of the resource stock, for example, the costs of ground water remediation.

direct approaches—valuation approaches that use survey-based techniques to elicit preferences for nonmarket goods and services (e.g., the contingent valuation method).

discounting—a procedure that adjusts for future values of a particular good by accounting for time preferences. It aims to determine the present value of benefits or costs in relation to the benefits or costs at different times in the future.

drawdown—lowering of the water table or potentiometric surface as a result of pumping.

existence value (EV)—a pure nonuse value that is the amount an individual would pay to know that a resource exists.

extractive value—a value calculated by adding up the benefits (across time) of removing water from an aquifer.

future value (FV)—the value of a time-dependent benefit or cost obtained by discounting the values out into future.

hedonic price method (HPM)—a technique to estimate implicit prices for environmental attributes of a market commodity. Usually applied to levels of clean air or water or similar environmental attributes as components of the total value (sales price) of real estate (Young, 1996).

incremental value—see **marginal value**.

indirect approaches—approaches that rely on observed behavior to infer nonmarket values. Examples include the travel cost method and the hedonic price method.

in situ—in place, that is, within the aquifer.

intergenerational equity—the concept that one generation should consume in a manner that allows an equal opportunity for future generations.

irreversibility—an effect that cannot be restored to its original state.

law of capture—as applied to common property resources is the tenet that whoever gets to the ground water first gets to use it.

marginal value—the value of another (hypothetical or last) increment of water when used in the most efficient manner.

nanofiltration—a membrane filtration process designed to desalinate (or soften) water at relatively low pressure.

natural discharge—ground water that reaches the surface (streams, lakes, wetlands, etc.) in the absence of pumping, excavation, or other human action.

neoclassical welfare economics—a school of economic thought in which the basic premises are that all economic activity is aimed at increasing the welfare of the individuals in society and that individuals are the best judges of their own welfare.

nonmarket—describes goods and services not priced and traded in markets.

nonmarket and market value—a market value is the value ascertained in a system where supply and demand forces are free to work and set the value of a good. A nonmarket value is the value of a good outside of a market system (for example, the value of the air you breathe or of the beautiful stream in the park).

nonpoint source pollution—pollution that is caused by diffuse sources that are not regulated as point sources and are normally associated with agricultural, silvicultural, and urban runoff.

nonuse value—values that are independent of the people's present use of the resource. (See bequest value and existence value.)

open access resource—a resource that has unlimited access which causes overexploitation of the resource.

option value—often categorized as a nonuse or passive use value, and refers to the fact that an individual places a certain current value on the option to use a resource in the future. Option value is often considered "closest" of the preservation values to use values.

overdrafting—ground water supply that is being used in excess of its natural recharge rate.

potentiometric surface—represents the height of rise of the water due to hydrostatic pressure when the constraint of the confining layer is removed (see Figure 2.2). Sometimes referred to as the piezometric surface.

present value—the value of a time-dependent benefit or cost obtained by discounting the values back through time to the present.

rate of time preference—rate of conversion of value between time periods. It is defined at the individual level; it is a feature of peoples' desires.

recharge—the replenishment of water beneath the earth's surface, usually through percolation through soils or connection to surface water.

recreational value—one of the use values provided by water. It is rooted indirectly to the value of ground water, which plays a role in providing surface water.

regulatory impact analysis (RIA)—a required analysis to be performed to determine the benefits and costs of a proposed regulation.

renovated ground water—ground water from which certain contaminants have been removed, synonymous with remediated ground water.

reverse osmosis—is a highly efficient removal process for inorganic ions, salts, some organic compounds, and in some designs, microbiological contaminants. Reverse osmosis resembles the membrane filtration process in that it involves the application of a high feed water pressure to force water through semipermeable membrane. In osmotic processes, water spontaneously passes through semipermeable membrane from a dilute solution to a concentrated solution in order to equilibrate concentrations. Reverse osmosis is produced by exerting enough pressure on a concentrated solution to reverse this flow and push the water from the concentrated solution to the more dilute one. The result is clear permeate water and a brackish reject concentrate (NRC, 1997).

riparian—associated with stream banks.

service flows—indirect and direct benefits to consumers attributable to a provision of services over time.

total economic value (TEV)—the sum of the extractive and *in situ* values. From the perspective of how to calculate the total economic value, it is the sum of use and nonuse values.

travel cost method—an indirect valuation method that values a recreation site by estimating the demand for access to the site; expenditures on the travel required to reach a recreational site are interpreted as a measure of willingness to pay for the recreational experience (Young, 1996).

unconfined aquifer—see **confined aquifer**.

use value—determined by a resource's or environmental asset's contribution to current production or consumption. Values may take the form of changes in income to producers (e.g., increases in agricultural output) or changes in services not usually traded in markets, such as recreation. Use values thus involve both marketed and nonmarketed commodities. Changes may also involve "intermediate" commodities that affect final consumption or production (e.g., natural filtration of polluted water or other ecosystem functions).

value—what one is willing to give up in order to obtain a good, service, experience, or state of nature. Economists try to measure this in monetary terms.

value taxonomies—a classification through which resource value or benefits reflect the economic channels through which a resource's service is valued (for example, **use** and **nonuse** values or **extractive** and ***in situ*** values).

welfare economics—a field of inquiry within the broad scope of economics that is concerned with money measures of individual and social well-being, particularly changes in well-being due to implementation of public policies.

willingness to accept (WTA) and **willingness to pay (WTP)**—willingness to accept is the minimum amount an individual must be paid to accept a certain risk or a change (decrement) in environmental quality. Willingness to pay is the maximum amount an individual would pay to obtain a change (increment) in environmental quality.

References

National Research Council. 1997. Safe Water From Every Tap. Washington, D.C.: National Academy Press.

Young, R. A. 1996. Measuring Economic Benefits for Water Investments and Policies. World Bank Technical Paper No. 338. Washington, D.C.: World Bank.

Acronyms

ABM—averting behavior method
BCA—benefit-cost analysis
BV—bequest value
CVM—contingent valuation method
EV—existence value
HPM—hedonic price method
RIA—regulatory impact analyses
TEV—total economic value
TCM—travel cost method
WTA—willingness to accept
WTP—willingness to pay

Appendix B

A Portion of a Sample Contingent Value Methods Questionnaire

John Reed Powell

1. There are many ways to protect public water supplies, and communities vary greatly in the extent to which they protect water supplies. We would like your opinion on how well you feel the drinking water supply in your community is protected from contamination.

Study the chart below and circle one letter (A, B, C, or D) that corresponds to how safe you feel about your household drinking water supply.

HOW SAFE I FEEL	**PROTECTION LEVEL**	**DESCRIPTION**
VERY SAFE	A	I feel absolutely secure. I have no worries about the safety of the community water supply at present. I am certain the level of protection is excellent and I cannot foresee any contamination occurring in the near future.
SAFE	B	I feel secure. I am confident the community water supply is safe at present. I am sure the level of protection is good and I am reasonably sure the water will not be contaminated in the near future.
SOMEWHAT SAFE	C	I feel apprehensive. I am unsure about the safety of the community water supply. I think the level of protection is adequate at present, but I am uneasy about the future. There is a possibility it could become contaminated in the near future.

UNSAFE	D	I feel troubled. I am anxious about the safety of the community water supply. I have doubts about the level of protection and I think it is very likely the water will become contaminated in the near future.

2. One way to prevent pollution of the supply is to establish an areawide special water protection district. This district would develop and implement pollution prevention policies specifically designed to suit your community's needs. Those in the district who would benefit from the increased protection would make an annual payment, which would be added to their water utility bill.

We are interested in discovering what you would be willing to pay, in higher water utility bills, per year to increase the level of protection for your community water supply. Take into account your household income and the fact that the money would have come from some part of your budget.

Using the payment card below, please indicate how much you are willing to pay, per year, to go from your present protection level, which you indicated on the chart in Question 1, to the highest level (For example, if you feel "SOME-WHAT SAFE" now, what would you pay to move to "VERY SAFE")? *Circle one dollar range as your annual payment.*

$ 0	$51-75	$201-225
$0-10	$76-100	$226-250
$11-20	$101-125	$251-275
$21-30	$126-150	$276-300
$31-40	$151-175	$301-325
$41-50	$176-200	$326-350

If you would be willing to pay more than $350, what is the maximum amount per year that would pay $____________.

3. If you answered "$0" to Question 2, please indicate which one of the following reasons best describes why you answered the way you did:

[ ] I AM ALREADY AT THE HIGHEST PROTECTION LEVEL.
[ ] I NEED MORE INFORMATION TO ANSWER THE QUESTION.
[ ] I DO NOT WANT TO PLACE A DOLLAR VALUE ON WATER SUPPLY PROTECTION.

[ ] I AM ALREADY PAYING ENOUGH.
[ ] OTHER (Please specify)________________________________

4. Many individuals and groups supply information about water quality. How much trust do you have in information from each of the following different sources? Circle one letter for each item.

	TRUST IN SOURCE			
	DO NOT TRUST AT ALL	**SOME TRUST**	**GREAT DEAL OF TRUST**	**UNSURE**
Business and Industry	a	b	c	d
Citizen Groups	a	b	c	d
Municipal Water Supplier	a	b	c	d
Farmers	a	b	c	d
State Agency Officials	a	b	c	d
Newspapers	a	b	c	d
Local Government Officials	a	b	c	d
Scientific Experts	a	b	c	d
Environmental Groups	a	b	c	d

APPENDIX C

Biographical Sketches of Committee Members

Larry W. Canter, who chaired the committee, is the Sun Company Chair of Ground Water Hydrology, and Director of the Environmental and Ground Water Institute at the University of Oklahoma, Norman, Oklahoma. He holds a B.S. in civil engineering from Vanderbilt University, an M.S. in sanitary engineering from the University of Illinois, and a Ph.D. in Environmental Health Engineering from the University of Texas, Austin, Texas. Over the last 30 years, Dr. Canter's research has addressed a wide range of technical and policy issues relating to ground water quality and quantity. He has published over 110 papers in refereed journals or conference proceedings and is author of over 135 research reports. Dr. Canter is a former member of the U.S. Army Corps of Engineers Environmental Advisory Board and the Governor of Oklahoma's Coordinating Committee on Water Resources Research. He served as associate editor of *Environmental Professional* and on the International Advisory Board of *EIA Review*. He was a principal contributor to the WSTB's 1989 colloquium on "Ground Water and Soil Contamination Remediation: Toward Compatible Science, Policy, and Public Perception."

Charles W. Abdalla is an associate professor in the Department of Agricultural Economics and Rural Sociology at Pennsylvania State University. He earned his B.S. in environmental resource management from Pennsylvania State University, and his M.A. in economics, M.S. in agricultural economics, and

Ph.D. in agricultural economics from Michigan State University. His research and extension programs address public choice about natural resources and the environment. His recent studies include the economic assessment of institutions for water management, measuring ground water values, and implications of industrialization of the U.S. food system for environmental policy design. He has written several book chapters, articles, and reviews, his work appearing in journals such as *American Journal of Agricultural Economics, Land Economics*, and *Water Resources Bulletin.* Also he served as guest editor of an award-winning issue of the *Journal of Soil and Water Conservation* entitled "Rural Groundwater Quality Management: Emerging Issues and Public Policies for the 1990s." Dr. Abdalla is the co-founder of the Pennsylvania Groundwater Policy Education Project and has worked with organizations at the national, state, and local levels to inform citizens and public officials and increase their involvement in water resources decision-making. He is a recipient of the Gilbert White Fellowship awarded by Resources for the Future and the Berg Fellowship awarded by the Soil and Water Conservation Society.

Richard M. Adams is a professor of agricultural and resource economics at Oregon State University, Corvallis. He holds a B.S. in resource management, and an M.S. and Ph.D. in agricultural economics from the University of California, Davis. His current research interests include economic assessments of the effects of environmental change on agricultural and natural ecosystems, assessments of agricultural costs and social benefits of reducing ground water pollution from agriculture, valuation of water across competing uses, and valuation of fish and wildlife resources. He has served as associate editor of *Water Resources Research*, editor of the *American Journal of Agricultural Economics*, and chair of the Policy Sciences Committee of the American Geophysical Union's Hydrology Section. Dr. Adams is author of approximately 200 publications and has provided consulting services on numerous natural resource assessment projects.

J. David Aiken is a professor of water and agricultural law at the University of Nebraska, Lincoln. He is a lawyer and has written extensively on ground water policy, including topics such as the depletion of the Ogallala aquifer, ground water protection, and conjunctive use. He has made extensive study of the various approaches to ground water allocation and protection taken in the western states. In addition, he has served as a legal consultant in the development of Nebraska's recharge appropriation legislation, and on a variety of agricultural and water law cases. He was a member of the Nebraska Water Independence Congress, and currently serves on the Water Quality Technical Advisory Committee of the Nebraska Department of Environmental Quality.

Sandra O. Archibald earned her B.A. and M.S. in Public Policy from the University of California at Berkeley, and her M.S. and Ph.D. in Agricultural Economics from the University of California at Davis. She is an associate professor of public affairs and planning at the University of Minnesota. Her research on the economic valuation of water resources has been conducted primarily in California where she heads a large study examining the impacts of changing federal water policy on ground water value. Her research has addressed issues such as ground water and surface water interactions, spatial and temporal dimensions, institutional roles, surrogate use indicators, and sustainability. In addition, she has examined economic issues related to irrigation drainage in the San Joaquin Valley, and the economic impacts associated with the ecological effects of ground water mining. Dr. Archibald has also published in the area of pesticide use and associated economic benefits and health risks. She is a member of the American Agricultural Economics Association, and has served on committees of the Transportation Research Board and Institute of Medicine.

Susan Capalbo is an associate professor in the Department of Agricultural Economics and Economics at Montana State University in Bozeman, Montana. She has also held positions at the University of California, Davis; Resources for the Future; the University of Maryland; and the National Marine Fisheries Service. Dr. Capalbo's current research examines the interface among agricultural practices and the environment, with a specific emphasis on ground water, surface water, erosion, and farm worker health. She has served as director of the Western Agricultural Economics Association and associate editor of the *American Journal of Agricultural Economics*. She is a member of the American Agricultural Economics Association, American Economic Association, Canadian Economic Association, Western Agricultural Economics Association, and Association of Environmental and Resource Economists. She recently won the Outstanding Journal Article Award in the *Northeastern Journal of Agricultural and Resource Economics.*

Patrick A. Domenico (a committee member through November 1995) is the David B. Harris Professor of Geology at Texas A&M University's College Station Campus where he specializes in teaching and research in ground water hydrology. He earned his B.S. in geology and M.S. in engineering geology from Syracuse University, and his Ph.D. in hydrology from the University of Nevada. Dr. Domenico has authored more than 40 professional publications, including a textbook on ground water hydrology. He has consulted on projects dealing with hydrologic, ground water supply, geothermal, and environmental issues for many private and governmental organizations including the International Bank for Reconstruction and Development, DuPont Chemical Company, and the Edison Electric Institute. He has received many prestigious awards, including the Birdsall Distinguished Lecturer in Hydrogeology, the Distinguished Teaching Award from the College of Geoscience, and the Distinguished Teaching Award from Texas A&M University.

Peter G. Hubbell has formed a private consulting firm, Water Resource Associates, Inc., focusing on water resource engineering and planning. He was formerly Executive Director of the Southwest Florida Water Management District (SWFWMD). He holds a B.S. in Water Resource Management from the University of Maryland, and completed the Program for Senior Executives at Harvard University's John F. Kennedy School of Government. At the SWFWMD he was responsible for the development of water resource programs and overall management of District operations. He has also held positions as a water resource analyst with a major environmental and engineering consulting firm, with the U.S. Bureau of Land Management, and with the U.S. Geological Survey. He was the chair of Florida's Bluebelt Commission on Aquifer Recharge, a founding member of the International Water Resource Network's Policy Council, a member of the board of the Florida Conflict Resolution Consortium, and holds memberships in the American Water Resources Association, American Water Works Association, and Interstate Council on Water Policy.

Katharine L. Jacobs is the Director of the Arizona Department of Water Resources Tucson Active Management Area. She holds a B.A. in Biology from Middlebury College, and an M.L.A. in environmental planning from the University of California, Berkeley. As Director of the Tucson Active Management Area, she is involved in ground water policy development and coordination, and consensus building in solving water resource management problems. She represents southern Arizona in statewide water issues and works with community leaders and other agencies to address a variety of local resource problems. She was a primary author in the development of Arizona's "assured and adequate water supply" rules. She serves on the Board of Directors of the Southern Arizona Water Resources Association and the Tucson Regional Water Council; and is a member of the Arizona Hydrologic Society, American Water Works Association, and American Water Resources Association.

Aaron L. Mills is a professor of environmental sciences at the University of Virginia. He earned his B.A. in biology from Ithaca College, and his M.S. and Ph.D. in soil science, with minors in microbiology and ecology, from Cornell University. Dr. Mills' areas of research address the transport of bacteria through porous media and biogeochemical reactions pertinent to ground water systems and he has over 80 professional publications. He has served as the chair of the aquatic and terrestrial microbiology section of the American Society for Microbiology and is an active member of the American Academy of Microbiology and the American Geophysical Union. He has also served as a member of the editorial board of *Applied and Environmental Microbiology* and *Microbial Ecology*. He has also served as a consultant on a number of sites where bioremediation efforts have been proposed and undertaken.

William R. Mills, Jr. has been the General Manager of the Orange County Water District (OCWD) since the fall of 1987. OCWD is responsible for management of the ground water basin in northern Orange County. Prior to his appointment at OCWD, Mills was a private consulting engineer between 1984 and 1987, specializing in water resources management, surface and ground water investigations, water quality, and water rights. From 1967 to 1984 Mills was employed by PRC Engineering, Inc. in Santa Ana, California in technical capacities up to and including President of the Planning and Development Division. Prior to joining PRC Engineering, Inc., Mills worked for the California Department of Water Resources and Los Angeles County Flood Control District. He is a graduate geological engineer from the Colorado School of Mines, with an M.S. degree in civil engineering from Loyola University of Los Angeles, and is a registered engineer and geologist.

Paul V. Roberts is the Clair Peck Professor of Environmental Engineering at Stanford University. He holds a B.S. in chemical engineering from Princeton University, a Ph.D. in chemical engineering from Cornell University, and an M.S. in environmental engineering from Stanford University. His research focuses on the transport and fate of contaminants in subsurface porous media, as well as water treatment technology. Previously, he headed the engineering department of the Swiss Federal Institute of Water Supply and Water Pollution Control. He has also worked as a research engineer at Stanford Research Institute, and as a process engineer at Chevron Research Company. He was a member of the WSTB's Committee on Ground Water Clean-Up Alternatives and has served three terms on the Environmental Engineering Committee of EPA's Science Advisory Board.

Thomas C. Schelling is a Distinguished Professor of Economics and Public Affairs at the University of Maryland, and the Lucius N. Littauer Professor of Political Economy, Emeritus, at Harvard University. He earned an A.B. in economics from the University of California, Berkeley, and a Ph.D. in economics from Harvard University. He has served in the U.S. Bureau of the Budget, the Economic Cooperation Administration in Europe, and the White House Executive Office of the President. He joined the faculty at Harvard University after serving five years on the faculty at Yale University. He is a member of the National Academy of Sciences, Institute of Medicine, and a fellow of the American Academy of Arts and Sciences, the American Association for the Advancement of Science, and the Association for Public Policy Analysis and Management. Dr. Schelling currently serves as a member of the NRC's Commission of Geosciences, Environment, and Resources.

Theodore Tomasi is a Research Scientist in the College of Marine Studies at the University of Delaware, and a Principal in the consulting firm Environmental Economics Research Group. He earned his B.A. in environment and public policy and an M.A. in economics from the University of Colorado, and a Ph.D. in natural resource economics from the University of Michigan. Dr. Tomasi's research focuses on environmental economics with special emphasis on methods for assessing the value of nonmarket goods and services. Prior to joining the University of Delaware, he held positions at the University of Minnesota, University of Michigan, and Michigan State University. He has served as a consultant on various issues of conducting natural resource damage assessments to NOAA, the U.S. Fish and Wildlife Service, the U.S. Department of Justice, the State of Michigan, and the State of Florida.

Index

E

F

G

H

I

K

L

M

N

O

P

Q

R